Monographs in Electrical and Electronic Engineering

Editors: P. Hammond and R. L. Grimsdale

Monographs in Electrical and Electronic Engineering

Stepping motors and their microprocessor controls

Takashi Kenjo

Professor in the Department of Electrical Engineering
and Power Electronics,
University of Industrial Technology,
Kanagawa, Japan

CLARENDON PRESS • OXFORD

Oxford University Press, Walton Street, Oxford OX2 6DP
Oxford New York Toronto
Delhi Bombay Calcutta Madras Karachi
Petaling Jaya Singapore Hong Kong Tokyo
Nairobi Dar es Salaam Cape Town
Melbourne Auckland
and associated companies in
Berlin Ibadan

Oxford is a trade mark of Oxford University Press

Published in the United States
by Oxford University Press, New York

First published 1984
Paperback edition 1985
Reprinted (with corrections) 1986, 1990

An earlier version of this work was published in Japanese in 1979
by Sogo Electronics Publishing Company

British Library Cataloguing in Publication Data
Kenjo, Takashi
    Stepping motors.—(Monographs in electrical
    and electronic engineering)
    1. Stepping motors
I. Title   II. Series
621.46'2   TK2511
ISBN 0-19-859339-2

Library of Congress Cataloging in Publication Data
Kenjo, Takashi.
    Stepping motors.
    (Monographs in electrical and electronic engineering)
    Includes bibliographies and index.
    1. Stepping motors. I. Title. II. Series.
TK2785.K4   1983   621.46'2   83-13193
ISBN 0-19-859339-2 (U.S.)

Set and printed in Northern Ireland at
The Universities Press (Belfast) Ltd.,

Preface

In the autumn of 1978, after I had completed the manuscript of the Japanese book *Fundamentals and applications of stepping motors* in collaboration with Y. Niimura, I visited Professor P. J. Lawrenson and Dr A. Hughes at the University of Leeds. I contributed one chapter of the book on the explanation of the theory of static and dynamic characteristics of stepping motors which had been developed by them and their co-workers. I told them that the small precision electrical motor industry in Japan had progressed at a spectacular pace since the early 1970s and that the book would be read by many engineers and students. Professor Lawrenson asked me if I had thought about writing its English version, and I replied 'I will consider it after the Japanese version has appeared'. It was this discussion that first inspired me to write this book.

When the Japanese version was published the next year, I sent several copies to specialists overseas. One of them arrived on the desk of Dr S. J. Yang of Heriot–Watt University. As he has some knowledge of the Japanese language, he could understand its outline. The copy of the book and his comments on it were handed over to Professor P. Hammond, the editor of the *Monographs in Electrical and Electronic Engineering*, and he felt that this kind of book was needed in English. At first he wanted me to produce a simple translation. But I thought I should write a new version, entirely revising the old one and including up-to-date material, aiming at a book which would attract the interest of various sorts of readers: from beginners to specialists in numerical control equipment. When I next visited England and talked with Professor Hammond, he agreed to my idea.

I am pleased to acknowledge that this book was completed with the support, help, and encouragement of many people. Professor Lawrenson and Dr Hughes granted me the use of their papers and the *Proceedings of the International Conference on Stepping Motors and Devices*. Moreover, Professor Lawrenson carefully read the typed draft and gave me helpful suggestions on how to improve some parts. Mr Y. Niimura, Sanyo Denki Co., Ltd., kindly agreed to the reproduction of useful illustrations from our previous book, and provided a new photograph. Mr J. P. Pawletko, IBM Corporation, replied to my enquiries about his achievements in the application of stepping motors in computer products, and prepared useful photographs of a linear stepping motor. Dr B. H. A. Goddijn, N. V. Philips' Gloeilampenfabrieken, granted me use of a picture of a motor in his paper, and replied to my enquiries about the type of machine. Mr M. L. Patterson, Hewlett Packard, provided a photograph of an XY

plotter and allowed me to reproduce an illustration from his report. Mr
P. B. Tennant, Moore Reed Co. Ltd., prepared photographs of a stepping
motor used in a floppy disc drive. Mr K. Egawa, MINEBEA Co. Ltd.,
produced excellent pictures of VR stepping motors for this book. Mr T.
Kojima, Fanuc Ltd., gave me photographs of historically interesting
stepping motors and numerically controlled machines using these. Profes-
sor M. R. Harris, the University of Newcastle, allowed me to reproduce a
graph from his paper and sent me the original. Mr L. W. Langley, Inland
Motor Specialty Products Division, gave me permission to reproduce two
illustrations from his paper. From Dr G. Singh, Exxon Office Systems
Company, I received permission to use some illustrations. A picture of a
Philips motor was prepared through Mr S. Itakura, Nihon Micro Motor
Co., Ltd. I found, in the *Proceedings of the Annual Symposium on
Incremental Motion Control Systems and Devices*, edited by B. C. Kuo,
University of Illinois at Urbana-Champaign, many papers which were
useful in writing this book. Photographs were provided by Nippon Elec-
tric Co. Ltd., Daini Seikosha Co. Ltd., Cambridge Instruments, and Ricoh
Co. Ltd.; and permission to reproduce illustrations from the *Proceedings
of IEE* was granted by the Institution of Electrical Engineers.

In closing, I should like to give my thanks to the above-mentioned
persons and organizations. My thanks are also extended to the staff of the
Clarendon Press who patiently corrected and refined my English.

T. K.

Kanagawa, Japan
May 1983

Contents

1. Introduction and historical survey

This book concerns itself with the electrical machines called stepping or step motors, covering their construction, principles, theory, driving techniques, and applications. To begin with we will survey the brief history and technological advancement of stepping motors. This survey covers the solenoid–ratchet stepping motor and electrohydraulic stepping motor; only small numbers of these are still manufactured nowadays. Various types of stepping motor are referred to in this chapter, but the structural details of the modern motors will be discussed in the next chapter.

1.1 What is a stepping motor and what are its basic characteristics?

Figure 1.1 illustrates the cross-sectional structure of a typical modern stepping motor; this is a so-called single-stack variable-reluctance motor. We shall first study how this machine works, using this figure. The stator core has six salient poles or teeth, while the rotor has four poles, both stator and rotor cores being of soft steel. Three sets of windings are arranged as shown in the figure. Each set has two coils connected in series. A set of windings is called a 'phase', and consequently this machine is a three-phase motor. Current is supplied from a DC power source to the windings via switches I, II, and III. In state (1), the winding of Phase I is supplied with current through switch I, or 'phase I is excited' in technical terms. The magnetic flux which occurs in the air-gap due to the excitation is indicated by arrows. In state (1), the two stator salient poles of phase I being excited are in alignment with two of the four rotor teeth. This is an equilibrium state in terms of dynamics. When switch II is closed to excite phase II in addition to phase I, magnetic flux is built up at the stator poles of phase II in the manner shown in state (2), and a counter-clockwise torque is created owing to 'tension' in the inclined magnetic field lines. The rotor will then, eventually, reach state (3).

Thus the rotor rotates through a fixed angle, which is termed the 'step angle', 15° in this case, as one switching operation is carried out. If switch I is now opened to de-energize phase I, the rotor will travel another 15° to reach state (4). The angular position of the rotor can thus be controlled in units of the step angle by a switching process. If the switching is carried out in sequence, the rotor will rotate with a stepped motion; the average speed can also be controlled by the switching process.

Nowadays, transistors are used as electronic switches for driving a stepping motor, and switching signals are generated by digital ICs or a

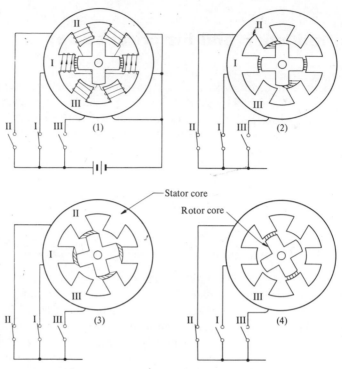

Fig. 1.1. Principle of a variable-reluctance stepping motor.

microprocessor (see Fig. 1.2). As explained above, the stepping motor is an electrical motor which converts a digital electric input into a mechanical motion. Compared with other devices that can perform the same or similar functions, a control system using a stepping motor has several significant advantages as follows:

(1) No feedback is normally required for either position control or speed control.

(2) Positional error is non-cumulative.

(3) Stepping motors are compatible with modern digital equipment.

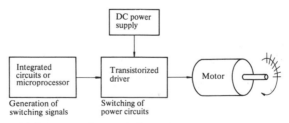

Fig. 1.2. Modern driving system for a stepping motor.

For these reasons, various types and classes of stepping motor have been used in computer peripherals and similar systems.

1.2 Early history of stepping motors

An issue of *JIEE*[1] published in 1927 carried an article 'The Application of Electricity in Warships', and a part of this article described a three-phase variable-reluctance stepping motor of the above type which was used to remote-control the direction indicator of torpedo tubes and guns in British warships. As illustrated in Fig. 1.3, a mechanical rotary switch was used for switching the excitating current. One revolution of the handle produces six stepping pulses causing 90° of rotor motion. The rotor motion in steps of 15° was geared down to attain the positional accuracy required.

It was pointed out in this article that in the design of this apparently simple stepping motor many factors have to be considered and many precautions taken in order to secure satisfactory operation. This machine requires a high ratio of torque to inertia of moving parts in order to avoid missing a step, and the time constant, the ratio of circuit inductance to resistance, should be small so as to attain a high speed of operation. These problems still apply to modern motors.

According to an article[2] in *IEEE Transactions on Automatic Control*, stepping motors were later employed in the US Navy for a similar purpose. Though practical applications of modern stepping motors occurred in the 1920s, the prototypes of variable-reluctance motors actually existed in earlier days. A paper by Byrne[3] says: 'Stepping type reluctance motors, now applied as positioning devices, were, as 'electromagnetic engines', the electric motors of the mid-nineteenth century.'

We shall here refer to two noteworthy inventions made in 1919 and 1920 in Britain.

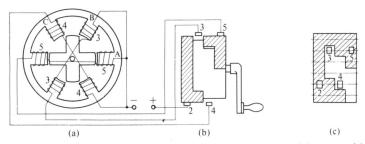

Fig. 1.3. A stepping motor used in British warships in the 1920s. (a) motor; (b) rotary switch; (c) stretched model of rotary switch.

(1) *Tooth structure to minimize step angle.* A UK patent[4] was obtained in 1919 by C. L. Walker, a civil engineer in Aberdeen, Scotland, for the invention of a stepping motor structure which can move in small step angles. Figure 1.4(a) and (b) shows respectively the longitudinal and cross-sectional views of a three-phase motor of the sort patented. The stator's salient poles each have a group of small teeth. The rotor teeth are of the same pitch as the stator's small teeth, the number of the rotor teeth being 32 in this figure. When phase I is excited and magnetic flux occurs along the path shown by the dotted curves in the figure, the groups of teeth in this phase are brought into alignment with some of the stator teeth as in Fig. 1.4(b). In this arrangement, the stator and rotor teeth in phases II and III must be out of alignment by 1/3 of the tooth pitch in opposing directions. When the excitating current is switched from phase I to phase II, the rotor will rotate clockwise through one step angle which is $(360/32)/3 = 3.75°$ in this case. However if the excitating current is switched to Phase III, the revolution will be counter-clockwise through the same small angle. Walker presented in the patent specification a plan for the construction of a type of stepping motor which is known nowadays as the multi-stack variable-reluctance type, as well as plans for the construction of a linear motor. It was, however, not until the 1950s that modern stepping motors employing the principle of this invention appeared commercially.

(2) *Production of a large torque from a sandwich structure.* C. B. Chicken and J. H. Thain in Newcastle upon Tyne in 1920 obtained a US patent[5] for the invention of a stepping motor which could produce a large torque per unit volume of rotor. The longitudinal construction of this machine is shown in Fig. 1.5(a), the remarkable feature of which is that the soft steel

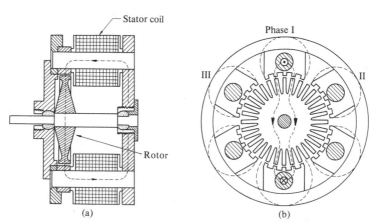

Stator coil

Phase I

III

II

Rotor

(a)

(b)

Fig. 1.4. A three-phase stepping motor invented by C. L. Walker.

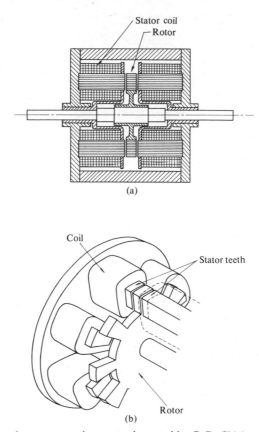

Fig. 1.5. A variable-reluctance stepping motor invented by C. B. Chicken and J. H. Thain:

rotor passes successively between the two opposite electromagnetic cores as shown in Fig. 1.5(b). This structure in which the rotor teeth are sandwiched by stator teeth is known to be the one which can produce the largest torque from a unit volume of rotor. But it was not until the 1970s that a stepping motor employing this principle was used as a power stepping motor in numerically controlled (NC) machines made by Fanuc Limited, a Japanese manufacturer.

1.3 Advent of the digital control age and progress in the 1960s

The January 1957 issue of *Control Engineering*[6] carried an epoch-making report on modern applications of stepping motors under the title 'The Power Stepping Motor—A New Digital Actuator'. This report concerned a system of three stepping motors which were applied to numerical three-axis contouring in a milling machine. The type of motor

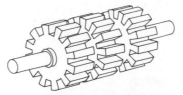

Fig. 1.6. Rotor of a multi-stack variable-reluctance stepping motor.

used in this system is the multi-stack variable-reluctance motor, whose rotor is as shown in Fig. 1.6; the switching devices were thyratron gas tubes. The driving system is shown in Fig. 1.7. The movements of the three motors are programmed manually or by means of a record in a punched film. The movements are read by a photoelectric reading head which supplies controlling signals to the thyratrons. The three stepping motors operate so as to cause the table to move in a three-axis space in a programmed order and so perform automatic machining.

Coinciding with the appearance of this report, intensive research work aimed at improving the performance of stepping motors was started in advanced industrial nations.

Since a large torque and output power are required for driving an NC machine, electrohydraulic stepping motors constructed by combining a regular stepping motor with an oil pressure mechanism were brought into use extensively in Japan in the period 1960 to 1974 (see Fig. 1.8). Figure

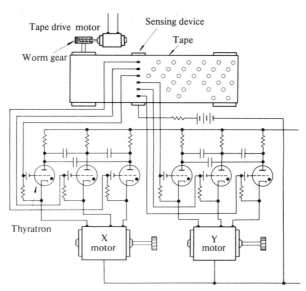

Fig. 1.7. Three-dimensional numerical control of a workpiece by means of variable-reluctance stepping motors driven by thyratrons. (After Ref. [6].)

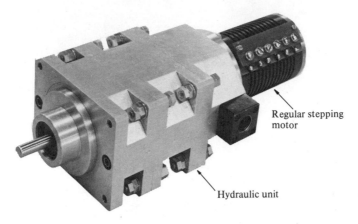

Regular stepping
motor

Hydraulic unit

Fig. 1.8. Electrohydraulic stepping motor. (By courtesy of Fanuc Ltd.)

1.9 is a milling machine, manufactured in 1961, which uses electrohydraulic stepping motors for three-axis operation, the switching elements
being germanium transistors.

From several reports ([7]–[11]), we know about the manufacture of
stepping motors in USA in the early 1960s. Twenty-eight manufacturers
are listed in Reference [10] and twenty-one in Reference [11], and it is
known that more than half of the manufacturers were building mechanical stepping motors or solenoid–ratchet motors. The construction and
mechanism of a typical solenoid–ratchet motor is given in Reference [8].

Three types of 'electromagnetic' stepping motor using permanent magnets had already appeared in those days, in addition to the variable-
reluctance motors. The simplest is the type which is now referred to as
the 'permanent magnet type' or 'PM motor'. The stator of this type of
motor has salient poles, while the rotor is a cylindrical permanent magnet
similar to that of a normal synchronous motor. The second type is a
hybrid motor, the rotor of which has the structure shown in Fig. 1.10; a
cylindrical permanent magnet polarized axially is covered by toothed soft
steel cores. This machine works as a stepping motor by the combined
principles of the permanent-magnet motor and the variable-reluctance
motor. This was invented by K. M. Feiertag and J. T. Donahoo of the
General Electric Company, and patented[12] in 1952 in USA. This type of
motor was manufactured first by the General Electric Company and
Superior Electric Company as a low-speed synchronous motor to run at
speeds as low as 100 r.p.m using a 60 Hz supply. It seems that the latter
company gave the product name 'Slo-syn' to this motor and gradually
improved it for use as a stepping motor. The third type of motor
employing permanent magnets is a single-phase stepping motor which has

Stepping motors

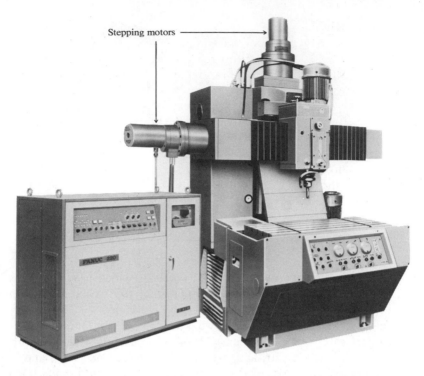

Fig. 1.9. A numerically controlled milling machine using electrohydraulic stepping motors. (By courtesy of Fanuc Ltd.)

been manufactured by Sigma Instruments, Inc. under the product name of 'Cyclonome' since 1952. This machine has a unique construction having two magnets in its stator, the form of which is illustrated in Fig. 2.71 on p. 61.

The size of stepping motors has since those days come to be referred to in the same way as servo-motor sizes (e.g. 08, 11, 15, 18, 20, 23, and 34 types where 08 and 11 refer to motors of diameter 0.8 and 1.1 inches respectively).

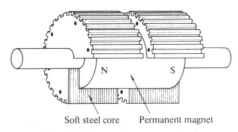

N   S

Soft steel core   Permanent magnet

Fig. 1.10. Rotor structure of the hybrid stepping motor.

From the beginning of the 1960s computer manufacturers took note of the possible uses of stepping motors as actuators in terminal devices and they promoted the development of reliable, high-performance motors. J. P. Pawletko, of the IBM Corporation, has been introducing stepping motors to many IBM products since the mid 1960s, and his activities have even extended to collaboration with motor manufacturers in designing machines. His first external paper[13] appeared in the 1972 Proceedings of the Annual Symposium on Incremental Motion Control Systems and Devices, which will be referred to later in this chapter.

In 1967, Sanyo Denki Co., Ltd. started serial production of hybrid motors under the product name of Step-Syn®. The Superior Electric Company, as stated earlier, was producing the Slo-syn® synchronous machines as a 1.8° stepping motor, and started full-scale serial production

Fig. 1.11. A large-capacity disc memory unit using ten stepping motors of 12° three-stack type. (By courtesy of Sanyo Denki Co., Ltd.)

Stepping motors are behind the side panel

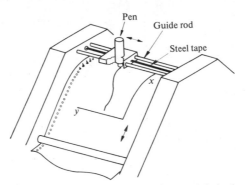

Fig. 1.12. Application of stepping motors to an XY-plotter made in the mid-1970s.

of hybrid motors in the 1970s as seen with the M-series. The production of 1.8-degree hybrid stepping motors by Sigma Instruments, Inc. started in 1969.

1.4 Rapid progress in the 1970s

Before the beginning of the 1970s, stepping motors with excellent dynamic performances became available. After entering the 1970s, a rapid increase was seen in the number of stepping motors used in the computer industry and this led to the mass production of the motors. In

USA, the manufacture of printers using stepping motors and DC servo motors started to emerge as a venture business. Printers adopting motor-control systems based on electronics, using integrated circuits instead of the complex mechanisms used in the past, must have had a strong appeal to young engineers because they permitted the free use of their knowledge of electronics. Similar trends were seen in other industrial nations in the early part of the 1970s. Salient types of stepping motors developed for applications to computer peripheral devices during the decade include:

four-phase motors of 1.8° steps;

four-phase hybrid motors of 2°, 2.5°, and 5° steps;

permanent-magnet motors of 7.5°, 45°, and 90° steps;

three- and four-phase variable-reluctance motors of 7.5° or 15° steps;

variable-reluctance motors of 128 or 132 steps per revolution. The last category covers single-duty motors for serial printers designed to match the number of characters per line used.

In the early part of the 1970s, automatic drawing machines[14] using a surface stepping motor of hybrid type appeared. It was in the latter part of the 1970s that linear motors of the variable-reluctance type started to be used for carriage transport in serial printers.[15],[16]

Progress was also made in the stepping motors used in NC machines. In 1973, Fanuc Ltd. was successful, under the leadership of S. Inaba, in the development of a unique high-power stepping motor. This is a multi-stack variable-reluctance motor, but it employs the sandwich structure suggested in Reference [2] to exert a high torque; it is used in numerically

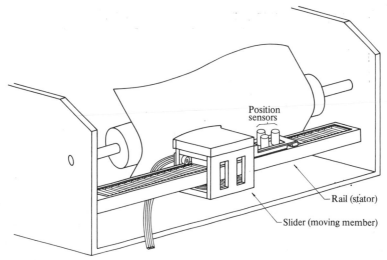

Fig. 1.13. Principle of a linear stepping motor applied to a serial printer. (After Ref. [15].)

controlled machines. It was, however, soon replaced by a DC servo-motor. One reason for this is the limitation that stepping motors have with regard to achieving a smooth finish, while another was the advancements made in the digital drive systems for DC motors.

In the field of computers, too, DC servo-motors have been used where high speeds and quick acceleration/deceleration are required, as in the drives of daisy wheels and capstans of magnetic tape handlers. But DC motors are subject to mechanical wear of brushes and commutator. Experiences with stepping motors show that they are free from mechanical wear problems and that they provide excellent reliability.

1.5 Advancement in semiconductor devices—enlarged application of stepping motors

As in other fields, technological progress associated with stepping motors took place hand-in-hand with the advance of transistors and other semiconductor devices. In 1948, the point-contact transistor was invented by the Bell Telephone Laboratory. Following the invention of junction transistors in 1950, solid state devices advanced at a truly spectacular pace. In 1957, the General Electric Company announced the development of the first thyristor under the product name of SCR (silicon controlled rectifier).

In the beginning, for stepping motor drives, mechanical contracts or relays were employed. Then, vacuum tubes or gas tubes were used, but they were gradually replaced by solid state devices such as thyristors and transistors. Although solid state devices came to offer greater ease of use year after year, the drive system including the logic circuitry was still expensive. So, while being viewed with high expectations, only a small

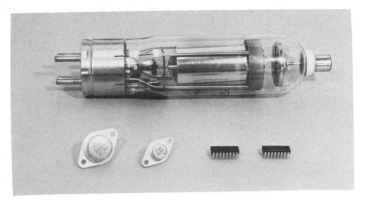

Fig. 1.14. Thyratron gas tubes, power transistors, and integrated circuits.

number of stepping motors were used in the 1960s. MOS transistors serviceable for practical use appeared in 1964. In the next year digital integrated circuits appeared. In subsequent years, development advanced to the stage of middle-scale integrated circuits, and further to large-scale integrated circuits. As a result, the logic circuit part in a drive system of a stepping motor was miniaturized, while ensuring increased reliability and lower costs. By this time, more applications of stepping motors became economically feasible. As mentioned above, it was during the 1970s that explosive growth was observed in the numbers of stepping motors used.

Advancement of semiconductor technology seems to have no end. In 1971, Intel Corporation announced the development of the four-bit microprocessor. Then the eight-bit microprocessor, having a larger range of application, was put on the market by Intel (1972) and Motorola (1974). As stated in the first part of this chapter, the stepping motor is an excellent machine when applied as an actuator in digital control systems. As microprocessors became available as the central processing unit, the realization of their application to the control systems for stepping motors seemed to be only a matter of time. In fact, in the latter part of the 1970s, microprocessors began to find various applications in stepping motor drives.

1.6 Academic activities

In the latter part of the 1950s intensive research on stepping motors was started in the universities and laboratories of advanced industrial nations.

Fig. 1.15. Microprocessor and peripheral devices.

The results of research were published in technical journals and magazines. In the 1970s, two international forums were established where specialists from the industrial and academic circles from all over the world gathered to exchange reports and hold discussions. One is the 'Symposium on Incremental Motion Control Systems and Devices' held annually since 1972 under the chairmanship of Professor B. C. Kuo of the University of Illinois at Urbana-Champaign, while the other is the 'International Conference on Stepping Motors and Devices' under the chairmanship of Professor P. J. Lawrenson of the University of Leeds, England, which has been held three times up to now (in 1974, 1976, and 1979). The stepping motor was also taken up as a subject in the 'International Conference on a Small Electrical Machines' held by the Institution of Electrical Engineers in London 1976.

References for Chapter 1

[1] McClelland, W. (1927). The Application of Electricity in Warships. *JIEE* **65**, 829–71. (Related part: pp. 850–2.)
[2] Kieburtz, R. B. (1964). The Step Motor—The Next Advance in Control Systems. *IEEE Transactions on Automatic Control*. January, pp. 98–104.
[3] Byrne, J. V. and Lacy, J. C. (1976). Characteristics of Saturated Stepper and Reluctance Motors. IEE Conf. Pub. **136** (Small Electrical Machines), pp. 93–6.
[4] Walker, C. L. (1919). Improvements in and connected with Electro-magnetic Step-by-step Signalling and Synchronous Rotation. UK Patent 137,150.
[5] Chicken, C. B. and Thain, J. H. (1920). Electrical Signaling Apparatus. US Patent 1,353,025.
[6] Thomas, A. G. and Fleischauer, F. J. (1957). The Power Stepping Motor—A New Digital Actuator. *Control Engineering* **4**, (Jan.), 74–81.
[7] Bailey, S. J. (1960). Incremental Servos, Part I—Stepping vs. Stepless Control. *Control Engineering* **7**, (Nov.) 123–7.
[8] Bailey, S. J. (1960). Incremental Servos, Part II—Operation and Analysis. Ibid. **7**, (Dec.) 97–102.
[9] Bailey, S. J. (1961). Incremental Servos, Part III—How They've Been Used. Ibid. **8** (Jan.), 85–8.
[10] Bailey, S. J. (1961). Incremental Servos, Part IV—Today's Hardware. Ibid. **8**, (March), 133–5.
[11] Proctor, J. (1963). Stepping Motors Move In. *Product Engineering* **4**, (Feb. 74–88.
[12] Feiertag, K. M. and Donahoo, J. T. (1952). Dynamoelectric Machine. US Patent 2,589,999.
[13] Pawletko, J. P. (1972). Approaches to Stepping Motor Controls. *Proc. First Annual Symposium on Incremental Motion Control Systems and Devices.* Department of Electrical Engineering, University of Illinois. pp. 431–63.
[14] Hinds, W. E. (1974). The Sawyer Linear Motor. *Proc. Third Annual Symposium on Incremental Motion Control Systems and Devices.* University of Illinous. pp. W1–W10.

[15] Chai, H. D. and Pawletko, J. P. Serial Printer with Linear Motor Drive. US Patent 4,044,881.

[16] Singh, G., Gerner, M., and Itzkowitz, H. (1979). Motion Control Aspects in the Qyx Intelligent Typewriter. *Proc. International Conference on Stepping Motors and Systems*. University of Leeds. pp. 6–12.

2. Outline of modern stepping motors

In the last chapter we followed the historical development of stepping motors in connection with numerical control technology. In this chapter, general views of the various types of stepping motors now used will be taken, and their structures and fundamental principles will be discussed without using mathematics. Technical terms which are used in association with stepping motors will be defined and their meanings studied.

2.1 Open-loop control systems

Generally, stepping motors are operated by electronic circuits, mostly on a DC power supply. The stepping motor is a unique motor in this respect as compared with conventional electric motors, which are mostly driven directly from a power supply. Moreover, stepping motors find their applications in speed and position controls without expensive feedback loops. This driving method is referred to as the open-loop drive. This section will deal with the fundamentals of the open-loop control of stepping motors. The details of electronic circuits for open-loop operation are not discussed here, but in Chapter 5.

Though open-loop control is an economically advantageous driving method, it is not free from some limitations. For example, the revolution of the rotor becomes oscillatory and unstable in certain speed ranges, and due to this behavioural characteristic, the speed and acceleration of a stepping motor controlled in the open-loop scheme cannot be as fast as a DC motor driven in a feedback-control scheme. Hence, in trying to expand the ranges of applications, the suppression of oscillations is a fundamental problem to be solved. Closed-loop control, a very effective driving method being free from instability and capable of quick acceleration, will be discussed in Chapter 7.

2.1.1 *System configuration*

To understand the fundamental configuration of the stepping motor driving system, let us examine the drive of a tape punch which employs a stepping motor to send the paper tape (Fig. 2.1). Operating instructions for numerically controlled machines are stored in the form of perforations made by this device. The tape drive system can be considered in the form of a block diagram shown in Fig. 2.2. As will be explained later, the stepping motor used for this purpose is usually a three- or four-phase motor. Here we examine a three-phase motor having three sets of winding.

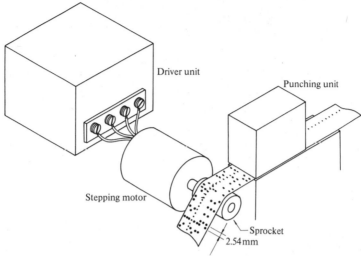

Fig. 2.1. Paper-tape perforator.

The most important feature of the stepping motor is that it revolves through a fixed angle for each pulse applied to the logic sequencer. The rated value of this angle (degrees) is referred to as the step angle.

Upon receiving a step command pulse, the logic sequencer determines the phase to be excited (or energized) and the phase to be de-energized, and it sends signals to the motor driver which is the stage which controls current supplied to the motor. The logic sequencer is usually assembled with TTL or CMOS integrated circuit chips. When the potential of an output channel from the logic sequencer is on level H (= high), the power driver works to excite the corresponding phase of the winding. Similarly, if the output is at level L, the phase of the same number is not excited, or it is turned off. As shown in Fig. 2.3, if the motor runs clockwise (= CW) in the excitation sequence of $1 \rightarrow 2 \rightarrow 3 \rightarrow 1 \cdots$, the direction of rotation will be counter-clockwise (= CCW) on the reversed sequence of $1 \rightarrow 3 \rightarrow 2 \rightarrow 1 \cdots$. In the tape punch system the sequence is usually fixed to send tape in one direction only. In general, there is no established rule to define which direction is clockwise or counter-clockwise, for a motor

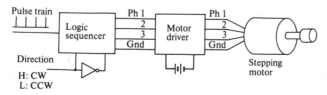

Fig. 2.2. Block diagram for motor drive system.

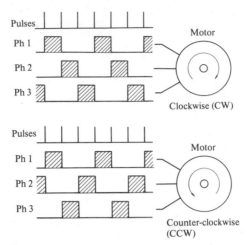

Fig. 2.3. Input pulse series and excitation sequence.

rotating clockwise when seen from one end appears to be rotating counter-clockwise if seen from the other end. The direction of rotation is usually defined by agreement between parties concerned.

In this book the phases are denoted by Ph1, Ph2, Ph3, etc. or PhA and PhB for some two-phase motors. The excitation used in Fig. 2.3 is referred to as the single-phase or one-phase excitation, which means that only one phase of the three (or four in a four-phase motor) is energized at any time. The single-phase excitation is often cited to serve in explanations of the fundamental principles of the stepping motor. But this is not always the best driving method. Details of this problem will be discussed in Section 2.3.

2.1.2   *Step and increment*

In recent numerically controlled machine systems and computer peripherals, the data are recorded in eight tracks on a tape; that is, there can be eight signal holes per line. In addition, between the third and fourth rows of signal holes is a train of guide holes with which the sprocket teeth couple. The lines are placed at 1/10 inch ($= 2.54$ mm) intervals. When storing data on a tape either by a manually operated tape punch or a computer, the tape is sent 2.54 mm, stopped for a line to be punched, and then advanced again through another 2.54 mm then stopped again, and so on. The simplest way to advance the tape through a line pitch is to apply a single pulse to the logic sequencer, so driving the motor through one step, turning the sprocket by an angle equivalent to one step, and thus driving the tape through 2.54 mm. But another method is to drive the stepping motor several step angles to advance the tape by one line. For

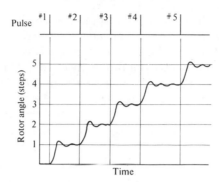

Fig. 2.4. Rotor angle and pulse series for low-frequency case.

example, a four-phase motor of 1.8° step angle could be employed to advance the tape one line pitch with four steps. The diameter of the sprocket is reduced to one quarter, and as a result, the inertia of the sprocket will be reduced to $(1/4)^4 = 1/256$ times that used in the one-step motion method.

The single motion which advances the paper tape through a line pitch, 2.54 mm in the above case, is often called one increment. One increment is realized by a single step in the former example, and by four steps in the latter example. In the system depicted in Fig. 2.2, the motor stops for a certain period of time after completing one increment of motion, allowing the tape to be punched, and this cycle repeats itself. This type of repeated starting and stopping motion is called 'incremental motion', and the control related to this type of operation is referred to as 'incremental-motion control'. Figures 2.4 and 2.5 show the overall relationship between the steps and the incremental operation. Figure 2.4 illustrates the relationship between the rotor angle and pulses applied to the logic sequencer, when it is assumed that the pulse frequency is relatively low, and it is the case that an increment of motion is performed with a single step. Figure 2.5 shows the cases in which one increment is performed with

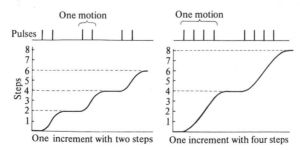

Fig. 2.5. Oscillation-free incremental motions with more than one pulse.

two or four steps. The single-step response is usually oscillatory as shown in Fig. 2.4. When a motion is carried out by several pulses at proper intervals, the response can be non-oscillatory. It is also possible to damp oscillation in the single-step operation mode by means of an electronic circuit technique. These matters will be discussed in Section 5.4.

The number of steps per increment is often more than four, for example in a tape-reader (Fig. 2.6). When data on a tape is transferred to the controller of a numerically controlled machine, the operation is implemented block by block. One block of data is composed of a number of lines or bytes, for example 32, 48, or 64 lines, and this number may differ depending on the system or case. Before a tool starts to move, a block of data is transferred to the semiconductor memory of the controller, and the tool is caused to move in the way that is instructed by the instructions in the first block of data. After those instructions are completed, the next block of data is read by the read head of the tape reader. If the system is designed to advance one line pitch in a single step, and the block is made up of 32 lines or bytes, one movement is made up of 32 steps. If one line pitch is advanced by four steps, one increment of motion is performed by 128 steps so as to transmit 32 bytes of data. Another stage must be placed before the logic sequencer if one motion or increment involves multiple steps. The stage for this function is termed

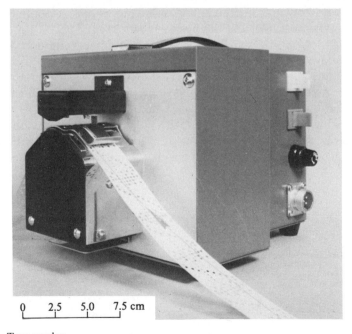

0   2.5   5.0   7.5 cm

Fig. 2.6. Tape-reader.

the 'input controller' in this book. The input controller sends out a train of a certain number of pulses at proper intervals, after receiving an input signal.

2.1.3 *Features of stepping motors from the viewpoint of application*

The features of the open-loop control of stepping motors are discussed, and some of the technical terms will be explained here.

(1) *Small step angle.* A stepping motor rotates through a fixed angle for each pulse. As explained earlier, the rated value of this angle is called the 'step angle' and is expressed in degrees. The smaller the step angle, the higher the resolution of positioning can be. One feature of stepping motors is that they can be made to realize a small step angle. Engineers are interested in the number of steps per revolution, which is conventionally termed the 'step number'. The relation between the step angle θ_s and the step number S is

$$S = 360/\theta_s. \tag{2.1}$$

Motors designed for use in the driving of character wheels (see Fig. 2.7) of a printer or typewriter are of 96, 128, or 132 steps per revolution. A standard four-phase motor has a step number of 200. Some precision motors are designed to attain one revolution with 500 or 1000 steps. However, the step angles in some simple motors are as large as 90°, 45°, or 15°.

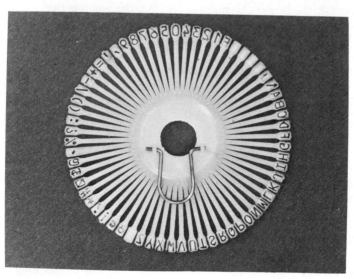

Fig. 2.7. Daisy wheel.

(2) *High positioning accuracy.* Accuracy in positioning is an important factor which determines the quality of a stepping motor. Stepping motors are designed so that they rotate through a predetermined step angle in response to a pulse signal and come to rest at a precise position. Since the accuracy at no-load depends on the physical accuracy of the rotor and stator, stepping motors are manufactured very carefully. Moreover, stepping motors are designed so that a high restoring torque is produced when a displacement from a rest position occurs due to a load torque. As will be discussed later on, the air-gap between the rotor and stator teeth is designed to be as small as possible to this end. Thus the positioning accuracy depends only on the machine characteristics and the driving circuit, while other electronic parameters have no effect on positioning accuracy.

Let us here consider some of the terminology involved in discussing the maximum static torque, the positions at which the rotor stops moving, and the accuracy in its positioning. Two concepts are defined for each as follows:

Maximum static torque[1]

(a) Holding torque: defined as the maximum static torque that can be applied to the shaft of an *excited* motor without causing continuous rotation.

(b) Detent torque: defined as the maximum static torque that can be applied to the shaft of an *unexcited* motor without causing continuous rotation.

In general, the larger the holding torque, the smaller the position error due to load (cf. Section 2.5.1). The detent torque appears only in motors having a permanent magnet.

Positions at which the rotor stops moving

(a) Rest position or equilibrium position: defined as 'the positions at which an excited motor comes to rest at no-load.

(b) Detent position: defined as the position at which a motor having a permanent magnet in its rotor or stator comes to rest without excitation at no-load.

In some motors the detent positions are utilized for positioning without exciting the windings so as to save power. The rest and detent positions are not always the same.

Positioning accuracy

(a) Step position error: defined as the largest positive or negative static angular position error (compared with the rated step angle) which can occur when the rotor moves from one rest position to the next.

(b) Positional accuracy: defined as the largest angular position error of a rest position related to the whole multiple of the rated step angle, which can occur during a full revolution of the rotor when moving from a reference rest position.

Examples of these definitions are given in Tables 2.1 and 2.2[2] on a 15° variable-reluctance motor, respectively for the positional accuracy and the step position error. As seen in Table 2.1 the errors are in the range from +0.08° to −0.03°. Consequently the positional accuracy is defined as 0.08° + 0.03° = 0.11°. When the errors are counted with reference to the third position as marked by * in the table, the maximum error will be +0.11°. This error is sometimes expressed as ±0.055°, for it might be possible to find a reference rest position at which the error ranges from +0.055° to −0.055°. As seen in Table 2.2, the error ranges from +0.11° to −0.09°. The step position error is defined as 0.11°.

Though both errors have the same value in this case by chance, the value of positional accuracy is usually larger than the step position error.

(3) *High torque-to-inertia ratio.* It is desirable that a stepping motor moves as fast as possible in response to an input pulse or pulse train. Not

Table 2.1. Example of data on positional accuracy measured on a 15° variable-reluctance motor. (After Ref. [2].)

Number of steps (n)	Theoretical angle ($n\theta_s$)	Measured angle (α_n)	Error ($\alpha_n - n\theta_s$)
0	0	0	0
1	15.00	15.06	+0.06
2*	30.00	29.97	−0.03
3	45.00	45.07	+0.07
4	60.00	60.00	0
5	75.00	75.06	+0.06
6	90.00	89.97	−0.03
7	105.00	105.07	+0.07
8	120.00	120.01	+0.01
9	135.00	135.05	+0.05
10	150.00	149.97	−0.03
11	165.00	165.07	+0.07
12	180.00	180.01	+0.01
13	195.00	195.05	+0.05
14	210.00	209.97	−0.03
15	225.00	225.08	+0.08
16	240.00	240.00	0
17	255.00	255.05	+0.05
18	270.00	269.97	−0.03
19	285.00	285.08	+0.08
20	300.00	300.00	0
21	315.00	315.05	+0.05
22	330.00	329.97	−0.03
23	345.00	345.07	+0.07
24	360.00	360.00	0

Table 2.2. Example of data on step position error measured on a 15° variable-reluctance motor. (After Ref. [2].)

Number of steps (n)	Measured step angle (β_n)	Error ($\beta_n - 15°$)
0	0	0
1	15.06	+0.06
2	14.91	−0.09
3	15.10	+0.10
4	14.93	−0.07
5	15.06	+0.06
6	14.91	−0.09
7	15.10	+0.10
8	14.94	−0.06
9	15.04	+0.04
10	14.92	−0.08
11	15.10	+0.10
12	14.94	−0.06
13	15.04	+0.04
14	14.92	−0.08
15	15.11	+0.11
16	14.92	−0.08
17	15.05	+0.05
18	14.92	−0.08
19	15.11	+0.11
20	14.92	−0.08
21	15.05	+0.05
22	14.92	−0.08
23	15.11	+0.11
24	14.93	−0.07

only a quick start but also a quick stop is required for a stepping motor. If the pulse train is interrupted while the motor is running at a uniform speed, the motor should be capable of stopping at the position specified by the last pulse. The above indicates that the ratio of the torque to rotor inertia must be large in stepping motors as compared with conventional electrical motors.

(4) *Stepping rate and pulse frequency.* The speed of rotation of a stepping motor is given in terms of the number of steps per second, and the term 'stepping rate' is often used to indicate speed. Since, in most stepping motors, the number of pulses applied to the logic sequencer equals the number of steps, the speed may be expressed in terms of pulse frequency. In this book, hertz (Hz) is used for the unit of stepping rate.

It should be noted, however, that the stepping rate does not specify the absolute speed. This speed is termed 'rotational speed' for conventional electrical machines and is expressed in terms of revolutions per minute; it is recommended that the same term and unit for the absolute speed of stepping motors be used. The relation between the rotational speed and

the stepping rate is given by

$$n = 60f/S \qquad (2.2)$$

where n = rotational speed (r.p.m.)

f = stepping rate (Hz)

S = step number.

2.2 Classification of stepping motors

A wide variety of electrical motors are in use, and the stepping motor can be classified into several types according to machine structure and principle of operation.

2.2.1 VR motor

The variable-reluctance stepping motor, or VR motor for short, may be considered to be the most basic type of stepping motor. The internal structure of a VR motor is illustrated in Fig. 2.8. The cross-sectional diagram of a simple motor in this category is illustrated in Fig. 2.9 to facilitate explanation of the basic motor principles. This is a three-phase motor having six stator teeth. Any two opposing stator teeth, which are at 180° from each other, belong to the same phase; that is, the coils on each opposing tooth are connected in series or parallel. (In the figure they are connected in series.) The rotor has four teeth. The stator and rotor core are normally made of laminated silicon steel, but solid silicon steel rotors are extensively employed. Both the stator and rotor materials must have high permeability and be capable of allowing a high magnetic flux to pass through them even if a low magnetomotive force is applied.

We shall see whether the two stator teeth in a phase should have the same magnetic polarity or opposite polarities to each other. Though this has a bearing on other problems, we assume in this example that the two

Fig. 2.8. Cutaway view of a single-stack VR motor. (By courtesy of MINEBEA Co., Ltd.)

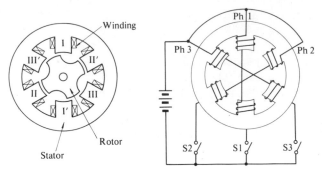

Fig. 2.9. Cross-sectional model of a three-phase VR stepping motor and winding arrangement.

teeth have opposite polarities. Hence, in Fig. 2.9, teeth I, II, and III form the north pole and teeth I', II', and III' the south pole when they are excited.

Current to each phase is controlled in the ON/OFF mode by their respective switches. If a current is applied to the coils of Ph1, or in other words if Ph1 is excited, the magnetic flux will occur as shown in Fig. 2.10. The rotor will then be positioned so that the stator teeth I and I' and any two of the rotor teeth are aligned. Thus when the rotor teeth and stator teeth are in alignment, the magnetic reluctance is minimized, and this state provides a rest or equilibrium position. If the rotor tends to move away from the equilibrium position due to some external torque applied to the rotor shaft, a restoring torque will be generated as shown in Fig. 2.11. In this figure the external torque is applied to turn the rotor clockwise (CW) and the rotor is displaced in the same direction. This will result in the curving of magnetic flux lines at the edges of the teeth of both the stator and rotor. Known as the Maxwell stress, magnetic lines of intensity have a strong tension, or in other words, magnetic lines have a tendency to become as short and straight as possible (like elastic strings). In Fig. 2.11 this effect is observed at the tooth edges, creating a counterclockwise torque to restore the rotor to alignment with the stator teeth.

Fig. 2.10. Equilibrium position with phase 1 excited.

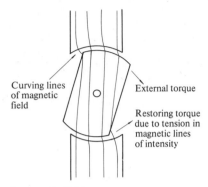

Fig. 2.11. Curving lines of magnetic field create torque.

As seen in the same figure, when the rotor and stator teeth are out of alignment in the excited phase, the magnetic reluctance is large. The VR motor works in such a way that the magnetic reluctance becomes a minimum. Let us see what will happen when Ph1 is turned off and Ph2 is turned on. The motor reluctance seen from the DC power supply will be suddenly increased just after the switching takes place. As is obvious from Fig. 2.12 the rotor will, then, move through a step angle of 30° counter-clockwise so as to minimize the reluctance. This motion through a step angle at each switching of excitation is called a step. After completing a rotor-tooth-pitch rotation in three steps, the rotor will apparently return to its original position. This is illustrated in Fig. 2.13.

Now we shall point out several basic structural features of the VR motors.

(a) *Air-gap should be as small as possible.* The air-gap between the stator teeth and rotor teeth in a stepping motor must be as small as possible to produce a high torque from a small volume of rotor and to attain high accuracy in positioning. Comparison between long and short gaps is given in Fig. 2.14. At the same level of magnetomotive force a small gap will

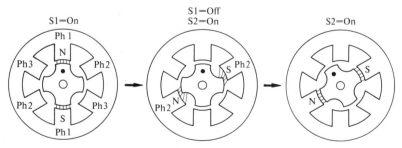

Fig. 2.12. How a step motion proceeds when excitation is switched from Ph1 to Ph2.

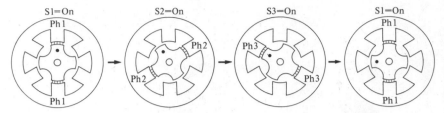

Fig. 2.13. Step motions as switching sequence proceeds in a three-phase VR motor.

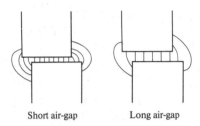

Short air-gap   Long air-gap

Fig. 2.14. Comparison of flux lines in a long and a short gap.

give more magnetic flux which produces a higher torque. It is also clear that the displacement from an equilibrium position is smaller with a smaller gap when an external torque is applied to the rotor. The gap size in modern motors is 30 to 100 μm.

(b) *For smaller step angles.* One of the unique features of the stepping motor is the possibility of realizing a small step angle. The 30° step angle realized by the structure in Fig. 2.10 is not a small angle. Figure 2.15(a) shows a three-phase motor with 12 stator teeth and 8 rotor teeth; this is twice as many as in the structure of Fig. 2.10. Figure 2.15(b) shows a four-phase motor having 8 stator teeth and 6 rotor teeth. The step angle

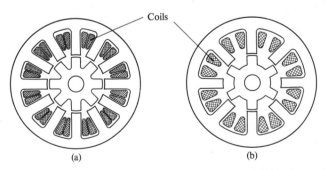

Coils

(a)   (b)

Fig. 2.15. Cross-sectional views of VR motors with 15° step angle. (a) Three-phase motor; number of stator teeth = 12; number of rotor teeth = 8. (b) Four-phase motor: number of stator teeth = 8; number of rotor teeth = 6.

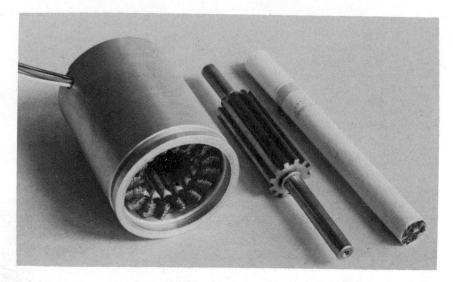

Fig. 2.16. Stator and rotor of a four-phase VR motor of 7.5° step angle. (By courtesy of Sanyo Denki Co., Ltd.)

in both structures is 15°. Figure 2.16 illustrates a four-phase 7.5° motor which has 16 stator teeth and 12 rotor teeth. As seen in this picture, the rotor of a stepping motor is very thin to minimize its moment of inertia.

The relationship between step angle θ_s, number of phases m, rotor teeth N_r, and step number S is given by

$$S = 360/\theta_s = mN_r.$$

For details refer to Section 3.3. In order to reduce the step angle θ_s the number of rotor teeth N_r must be increased. It may be taken from the above explanation that the number of stator teeth must be increased as

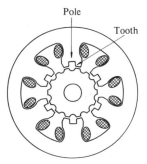

Pole

Tooth

Fig. 2.17. Cross-sectional view of a three-phase VR motor having two teeth on each pole; number of rotor teeth is 14 and step angle 8.75°.

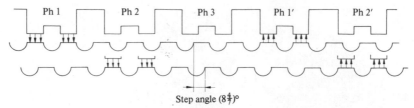

Step angle $(8\frac{4}{7})°$

Fig. 2.18. Split-and-unrolled model of the three-phase VR motor of the previous figure.

well as the rotor teeth. But, the number of stator teeth is not specified in the above equation. Actually, the cross-sectional diagram of a VR stepping motor with a small step angle is such as shown in Fig. 2.17. The large salient portions around which windings are located are conventionally called the poles. A pole has two or more stator teeth, and all the teeth in a pole have the same magnetic polarity at any time. Since the number of rotor teeth N_r is 14 and the number of phases is 3 in the structure of Fig. 2.17, the step number is $S = 3 \times 14 = 42$ steps. The step angle θ_s is $(360/42)° = 8.57°$ in this model. How this structure makes a step is shown in the split and unrolled model in Fig. 2.18. This example shows that the number of stator teeth is not a direct factor for determining the step angle.

An example in which the number of rotor teeth is increased to 44 in a three-phase motor is given in Fig. 2.19. The number of steps per revolution is 132. An example of a four-phase motor with 50 rotor teeth is given in Fig. 2.20. The step angle is 1.8° and the number of steps per revolution is 200 in this model. Figure 2.21 shows a series of six-phase VR motors which have a step angle of 1.2°.

(c) *Multi-stack type and single-stack type.* The VR stepping motors described above are all single-stack type motors. An outstanding feature of this type of motor is that three or four phases are arranged in a single stack, i.e. in the same plane. Another type of VR stepping motors is the

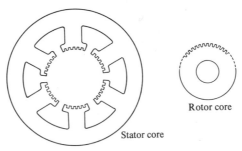

Rotor core

Stator core

Fig. 2.19. Cross-sectional view of a three-phase VR motor; number of rotor teeth being 44 and step number 132.

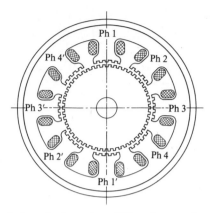

Fig. 2.20. Cross-sectional view of a four-phase VR motor; number of rotor teeth being 50, step number 200, and step angle 1.8°.

multi-stack type. This type is known also as the cascade type. A cutaway view of a three-stack motor is shown in Fig. 2.22. In this model each stack corresponds to a phase, and the stator and rotor have the same tooth pitch. Now we assume that the third phase (stack) is excited and the stator and rotor teeth are in alignment in this phase. In the other phases or stacks at this moment, the teeth of both members are misaligned by 1/3 tooth pitch, which is illustrated in the figure. The directions of misalignment are opposite to each other in the second and third stacks. If the excitation is switched from the third to the first phase, the rotor will move by one step in the CW direction seen from left, but if the excitation

Fig. 2.21. 1.2°, six-phase VR motors. (By courtesy of Nippon Pulse Motor Inc.)

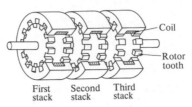

Fig. 2.22. Construction of a multi-stack VR motor.

is switched to the second phase the rotor will move in the CCW direction by one step angle. Figure 2.23 shows the stator and rotor of a five-stack motor.

Figure 2.24(a) is an axial diagram of a unique five-phase VR motor of multi-stack type manufactured by Fanuc Ltd and used in numerically controlled machines. Figure 2.24(b) shows the principle of this machine; the rotor teeth are sandwiched by stator teeth of the same tooth pitch. This structure is known to produce a large torque from a unit volume of rotor and to greatly improve the machine efficiency.

Since solenoid coils are used in the multi-stack motor shown here, the magnetic field distribution is different from the examples which have been presented so far in this book. The comparison is made in Fig. 2.25. In a single-stack motor, magnetic poles of even number (N, S, N, S · · ·) will appear in a plane perpendicular to the motor shaft. This type of magnetic field is called a heteropolar field. On the other hand only one magnetic

Fig. 2.23. Stator and rotor of a five-stack VR motor. (By courtesy of MINEBEA Co., Ltd.)

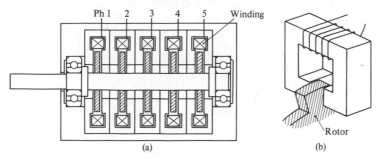

Fig. 2.24. Axial diagram of five-stack VR motor of sandwich type.

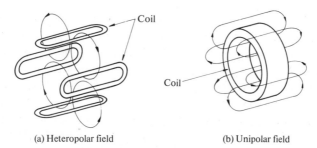

(a) Heteropolar field   (b) Unipolar field

Fig. 2.25. Heteropolar and unipolar magnetic fields.

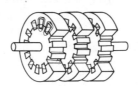

Fig. 2.26. A multi-stack VR motor of heteropolar field type.

pole (N or S) appears in a plane perpendicular to the shaft of a multi-stack motor. This type of distribution is unipolar. In some types of multi-stack VR motor the field distribution is heteropolar. An example is illustrated in Fig. 2.26.

2.2.2 *PM stepping motors*

A stepping motor using a permanent magnet in the rotor is called a permanent magnet (PM) motor. An example of a basic four-phase PM stepping motor is shown in Fig. 2.27. A cylindrical permanent magnet is employed as the rotor, and the stator has four teeth around which coils are wound. The basic scheme for the driving circuit is shown in Fig. 2.28. The terminal marked C at each of the phases is connected in turn to the

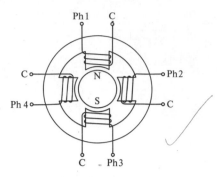

Fig. 2.27. Cross-sectional model of a four-phase PM motor.

positive terminal of the power supply. If the phases are excited in the sequence Ph1 → 2 → 3 ··· the rotor will be driven clockwise as shown in Fig. 2.29. The step angle is obviously 90° in this machine. If the numbers of stator teeth and magnetic poles on the rotor are both doubled, a four-phase motor with a step angle of 45° will be realized (see Fig. 2.30).

To decrease the step angle further, in a PM motor, the number of magnetic poles and stator teeth must be increased. However, there is a limit to the number of both teeth and magnetic poles. As an alternative the hybrid structure to be discussed in the next paragraph is widely employed in PM motors having a small step angle.

A feature of the PM motor is that the rotor comes to rest at a fixed position even if excitation ceases. This mechanism is referred to as the 'detent mechanism', and the predetermined positions as 'detent positions'. In general, the detent positions coincide with the excited rest positions (or equilibrium positions), as long as only one phase is excited. There are two disadvantages in the use of a permanent magnet; (1) the magnet is costly; and (2) the maximum flux density level is limited by the level of magnetic remanence of the magnet. Though ferrite magnet is cheap, it does not produce a high torque because of its inherently low remanence.

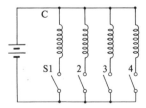

Fig. 2.28. Basic drive circuit for a four-phase motor.

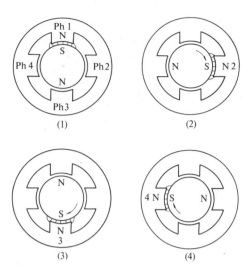

Fig. 2.29. Steps in a four-phase PM motor.

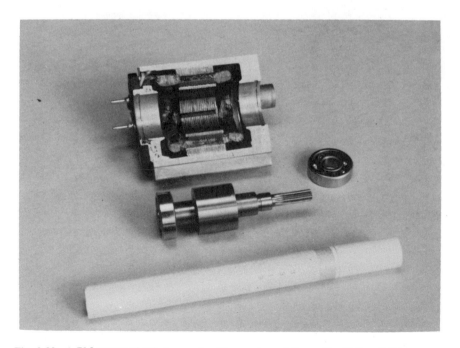

Fig. 2.30. A PM motor of 45° step angle. (By courtesy of Sanyo Denki Co., Ltd.)

2.2.3 *Hybrid stepping motors*

Another type of stepping motor having a permanent magnet in its rotor is the hybrid motor. The term 'hybrid' derives from the fact that the motor is operated under the combined principles of the permanent magnet and variable-reluctance motors. The cutaway view and the cross-sectional/axial diagram of the hybrid motor in wide use today are shown in Figs. 2.31 and 2.32, respectively. The stator core structure is the same as, or very close to, that of the VR motor shown in Fig. 2.20, but the windings and coil connections are different from those in the VR motor. In the VR motor only one of the two coils of one phase is wound on one pole, while in the four-phase hybrid motor, coils of two different phases are wound on the same pole as shown in Fig. 2.32. Therefore one pole does not belong to only one phase. The two coils at a pole are wound in the so-called bifilar scheme that will be discussed in Section 2.2.6; they produce different magnetic polarities on excitation.

Another important feature of the hybrid motor is its rotor structure. A cylindrically shaped magnet lies in the core of the rotor as shown in Fig. 2.33, and it is magnetized lengthwise to produce a unipolar field as shown in Fig. 2.34(a). Each pole of the magnet is covered with uniformly

Fig. 2.31. Cutaway view of a hybrid motor. (By courtesy of Sanyo Denki Co., Ltd.)

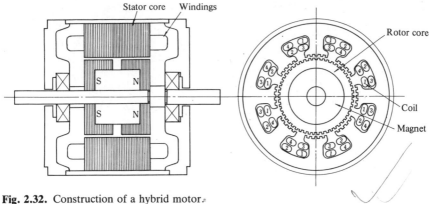

Fig. 2.32. Construction of a hybrid motor.

toothed soft steel. The teeth on the two sections are misaligned with respect to each other by a half tooth pitch. In some motors, the rotor teeth are aligned with each other, but the stator core has a misalignment as shown in Fig. 2.35. The toothed sections are normally made of laminated silicon steel, but sintered steel or solid silicon steel is employed in some cases.

The magnetic field generated by the stator coils is a heteropolar field as shown in Fig. 2.34(b).

In this type of motor, torque is created by the interaction of these two types of magnetic fields in the toothed structure in the air-gaps. For an explanation let us have a look at the split and unrolled model of Fig. 2.36. In this schematic diagram, the stator tooth pitch is the same as the rotor tooth pitch. In some motors, however, the stator tooth pitch is a little larger than the rotor's to reduce detent torque and increase positioning accuracy (see Fig. 3.16, p. 85). The upper half of this figure is the cross-sectional view of the south-pole side of the magnet while the lower half represents the north-pole side. We are concerned with the magnetic fields under the teeth of poles I and III in this model. Pole I is now

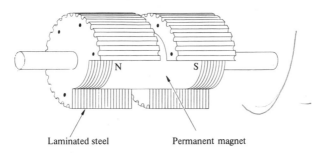

Fig. 2.33. Rotor structure of a hybrid motor.

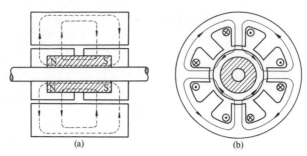

(a)   (b)

Fig. 2.34. Magnetic paths in a hybrid motor; (a) the. flux due to the rotor's magnet producing a unipolar field, while (b) shows the heteropolar distributed flux due to the stator currents.

excited to produce the north pole, and pole III the south pole and they build field distributions as drawn by the solid curves. The dotted curves represent the flux due to the magnet.

Firstly it should be noted that no effective torque is generated by the magnetic field due to the coil alone as in the VR motor, because the rotor teeth in the north-pole side and those in the south-pole side are misaligned with respect to each other by half a tooth pitch. The permanent magnet produces some detent torque, but this is not a very important factor in the hybrid motor. Let us see what happens when the magnetic fields due to the coils and the permanent magnet are superimposed. The results are suggested in the same figure. A driving force toward the left (←) will appear in the upper half section because both fields reinforce each other in the toothed structure under pole I so increasing the left-oriented force, while both components neutralize each other and so weaken the right-oriented force under pole III. The same force is produced as well in the lower half section, as the stator field and rotor field are in the same direction under pole III, while they are in opposing directions under pole I. Hence, the resultant force will be toward the left (←). After the rotor has moved a quarter-tooth pitch in this direction, the driving force is reduced to zero, and an equilibrium position is reached.

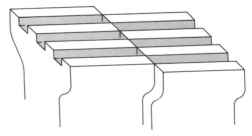

Fig. 2.35. Misalignment in stator teeth.

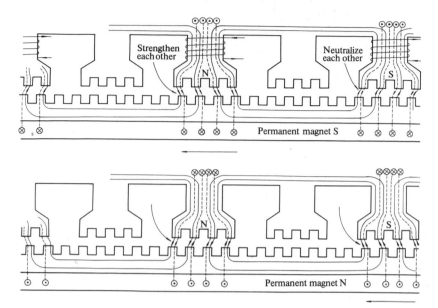

Fig. 2.36. Split-and-unrolled model of a four-phase hybrid motor; the upper being for the south-pole cross-section, and the lower the north-pole cross-section.

If the old excitation is turned off, and new poles are excited to produce a south pole and a north pole, respectively, the rotor will make another step. As seen above, the permanent magnet plays an important role in creating the driving force. But it should be noted that the toothed structures in both stator and rotor are designed to realize a small step angle in the hybrid stepping motor.

This type of motor was originally invented by Feiertag and Donahoo[5] and was designed to be used as a synchronous motor for low-speed application, being called a synchronous inductor motor. Actually some of today's hybrid motors can be used as two-phase or single-phase condenser-run synchronous motors. A 1.8° motor runs at 60 (or 72) r.p.m. on a 50 (or 60) Hz power supply. In these motors the windings need not be bifilar, the details of which will be discussed in Section 2.2.8.

The most popular hybrid motor is the four-phase 200-step motor, the step angle being 1.8°. This type of motor is produced by a number of manufacturers. There are, of course, hybrid motors with other step angles such as 2° or 5°. A German manufacturer, Gerhard Berger GmbH, produces five-phase hybrid motors, the details of which are found in References [6] and [7].

In order to raise the torque, multi-stack hybrid motors such as shown in Fig. 2.37 are employed.

Fig. 2.37. Three-stack hybrid motor designed to increase torque. (By courtesy of Sanyo Denki Co., Ltd.)

2.2.4 *Hybrid stepping motors having a permanent magnet in the stator*

Another type of hybrid stepping motor[8] is shown in Fig. 2.38. As indicated in its axial diagram in Fig. 2.39 the ring permanent magnet is placed in the stator, and the two windings are toroidal coils in this machine structure. Both the magnet and stator currents produce unipolar magnetic fields in this motor. The stator teeth in the two sections in a phase are misaligned with respect to each other by a quarter tooth pitch, while all the teeth in the four sections of the rotor are aligned with each other.

The driving principle of this machine is similar to that for the hybrid linear motor to be discussed in detail in Section 2.2.7.

2.2.5 *Permanent-magnet motor with claw-poles*

The claw-poled motor is another type of permanent-magnet stepping motor. The cutaway view of this machine is illustrated in Fig. 2.40. This is known also as the can-stack stepping motor. Since the stator of this motor consists of a sort of metal can. Teeth are punched out of a circular metal

Fig. 2.38. Stator and rotor of a hybrid ring-coil motor. (By courtesy of N. V. Philips' Gloeilampenfabrieken.)

sheet and the circle is then drawn into a bell shape. The teeth are then drawn inside to form claw-poles. A stack of the stator is formed by joining two bell-shaped casings so that the teeth of both of them are intermeshed and the solenoid coil is contained within them.

The feature of the can-stack motor is that the stator teeth or claw-poles produce a heteropolar magnetic field from a current flowing in the

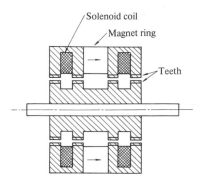

Fig. 2.39. Axial view of hybrid motor with a permanent magnet in its stator.

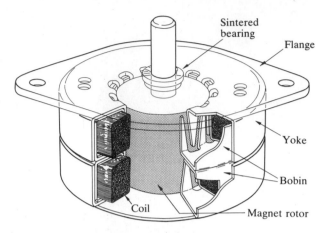

Fig. 2.40. Cutaway diagram of a claw-poled PM motor.

solenoid of the stator. As illustrated in Fig. 2.41, the rotor's cylindrical ceramic magnet (ferrite) is magnetized to produce a heteropolar field as well. A typical number of poles is 24 for a 7.5° step motor. This type of motor is usually of two stacks. When the rotor magnetization pitches under both stator stacks are aligned with each other as in Fig. 2.41, the stator tooth pitches in both stacks are misaligned by a quarter pitch.

The windings are either on a two-phase or a four-phase scheme. In the latter scheme, the windings of Ph1 and Ph3 are wound in the bifilar manner which will be discussed in Section 2.2.8, and placed in stack A, and those of Ph2 and Ph4 are in stack B. Ph1 and Ph3 are connected to produce opposing signs of magnetic polarity, and so are Ph2 and Ph4. The sequences of excitation are illustrated in the form of current graphs for both the four-phase and two-phase schemes in Fig. 2.42. In the two-phase scheme the applied current is an alternating square wave. We will examine the principles of this motor when based on the two-phase arrangement, using Fig. 2.43. Let us examine the positional relation of the rotor's magnetic poles and the stator teeth in stack A. The rotor is first placed in the position of state (1), and PhA is excited with a positive current so as to produce magnetic poles in the pattern given in the same figure. As is obvious, the rotor moves left by the tension of magnetic lines.

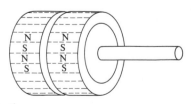

Fig. 2.41. Rotor magnetization

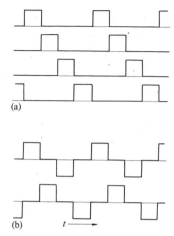

(a)

(b) $t \longrightarrow$

Fig. 2.42. Current waveform supplied to a claw-poled PM motor; (a) four-phase scheme; (b) two-phase scheme.

State (2) is an equilibrium position with PhA excited in the positive polarity. Next, if PhA is turned off and PhB is excited with a positive current, the rotor will be driven further in the same direction, because the stator teeth in stack B are misaligned by a quarter tooth pitch to the left with respect to the teeth in stack A. State (3) shows the resultant due to this excitation. To advance the rotor further to the left and place it in the next rest state (4), PhB is de-energized and PhA is excited with a negative current.

The claw-pole motor features low manufacturing cost and has many applications such as the paper feed prime mover and head drive motor in mini-size floppy disc drives.

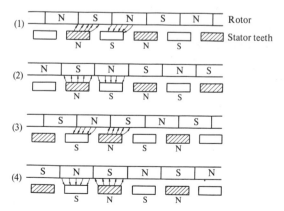

Fig. 2.43. How steps occur in a claw-poled PM motor.

2.2.6 *Outer-rotor stepping motor*

Rotating motors may be classified into the outer-rotor type and inner-rotor type. All the stepping motors described so far are the inner-rotor type, in which the stator encloses the rotor. In the outer-rotor stepping motor, the rotor is on the outside of the stator. Outer-rotor stepping motors are very rare, but exist.[9] The motor shown in Fig. 2.44 is an 81-step outer-rotor VR motor designed for a paper feeding system, the rotor itself being the roller for paper feed. As is obvious from the cross-sectional view of Fig. 2.45, the stator is of a three-stack type. The three stator cores are fastened to the shaft which is firmly attached to the device. The outer-rotor is connected to the shaft via two ball-races and is free to rotate.

2.2.7 *Linear stepping motors*

All the motors described so far are rotating machines. They are designed so that the rotor can rotate in both CW and CCW directions with reference to the stator.

There are, however, some motors which are designed to perform linear motion. They are called linear motors. There are as many kinds of linear motor as there are rotary motors; these include DC motors, synchronous motors, induction motors, and brushless motors. But the linear stepping motor is the most important one among the small linear motors used for control applications. Linear stepping motors may be classified into either

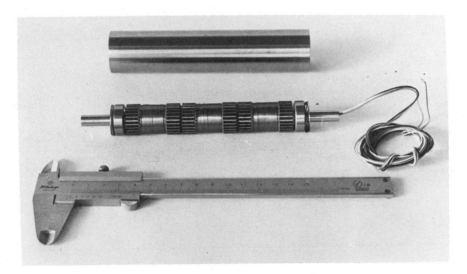

Fig. 2.44. An outer-rotor VR motor. (By courtesy of Sanyo Denki Co., Ltd.)

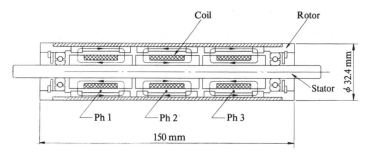

Fig. 2.45. Axial diagram of the outer-rotor motor shown in Fig. 2.44. (By courtesy of Sanyo Denki Co., Ltd.)

VR motors or PM motors, the latter corresponding to the hybrid motor of rotary type.

(1) *VR linear motor.* An example of the three-phase VR motor structure is shown in Fig. 2.46, which is a motor developed by IBM Corporation for transporting a carriage in a serial printer. Figure 2.47 shows the relation between the stator teeth, slider teeth, and windings. Both stator and slider cores are of laminated steel. The dynamic responses and control aspects of this sort of motor are discussed in Reference [10].

(2) *PM linear motors.* The principle of a PM motor known as the Sawyer linear motor[11] is illustrated in Fig. 2.48. The motor, which is here

Fig. 2.46. A three-phase linear VR motor used in a printer. (By courtesy of IBM Corporation.)

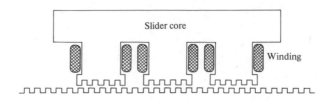

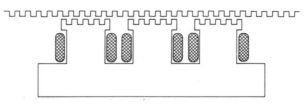

Fig. 2.47. Relation between the stator teeth, slider teeth, and windings.

termed 'slider' consists of a permanent magnet and two electromagnets, A and B. The flux due to the permanent magnet closes its path through the electromagnet cores, the air-gaps between core and stator and stator and core. In the absence of currents in the coil, the magnet flux flows through both core teeth as shown in electromagnet B in state (a) or (c). When the coil is excited, however, the flux concentrates into one tooth as shown in electromagnet A in (a). This brings the flux density in this tooth to a maximum, while that in the other tooth is reduced to a negligible value.

Now in Fig. 2.48(a) tooth 1 of electromagnet A is aligned with a stator tooth. When current is switched to coil B in the direction shown in (b), the slider will be driven right a quarter tooth pitch to bring tooth 4 in alignment with the adjacent stator tooth. Next, electromagnet B is de-energized and A is excited in the opposing polarity to before. This produces a force to bring tooth 2 in A in alignment with its adjacent stator tooth as shown in (c). To move the slider further in the same direction, coil A is de-energized and coil B is excited in the polarity opposite to before. This is the state in (d).

In the international conference on stepping motors and systems in 1979, a linear motor[12] which moves on a rod stator was presented. The basic machine principle is the same as the Sawyer motor as indicated by the construction in Fig. 2.49. The magnet used is of samarium–cobalt to minimize the slider mass.

The structure of a linear PM motor which was developed from the Sawyer principle to apply to the driving of a draft head in an automated

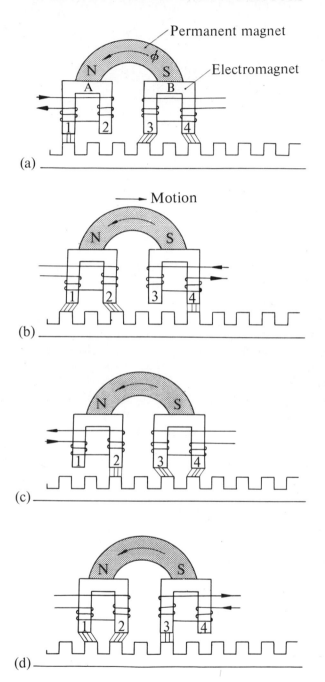

Fig. 2.48. Principle of the Sawyer linear motor.

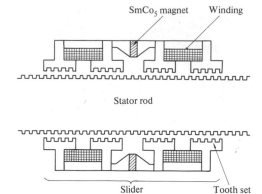

Fig. 2.49. Axial diagram of a PM linear motor. (After Reference [12], by courtesy of Inland Motor Specialty Product Division.)

drafting system is such as shown in Fig. 2.50. In this motor four flat magnets of samarium–cobalt are used.

2.2.8 *Monofilar and bifilar windings*

Now we are at the stage of discussing the types of windings used in stepping motors. Solenoid coils are used in multi-stack VR motors and claw-poled PM motors. On the other hand, for hybrid motors and single-stack VR motors, the monofilar winding or bifilar winding is used. In the former a single wire is wound a number of times on a single pole. In the latter, two overlapping wires are wound as one wire as shown in Fig. 2.51, and these two wires are separated from each other at the terminals to keep them as independent wires. If one coil belongs to Ph1, the other belongs to Ph3. Likewise, if one belongs to Ph2, the other belongs to Ph4.

One of the purposes of the bifilar winding is to energize a stator pole by alternating magnetic polarity. Excitation of a phase may be performed by

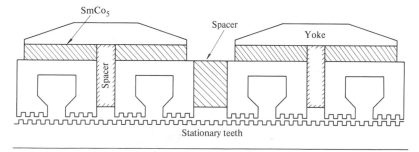

Fig. 2.50. Structure of a linear PM motor used in automated drafting machine.

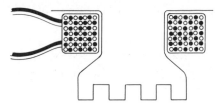

Fig. 2.51. Bifilar winding.

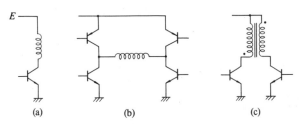

(a)   (b)   (c)

Fig. 2.52. Three fundamental exciting circuits: (a) monofilar winding, unipolar excitation; (b) monofilar winding, bipolar excitation; (c) bifilar winding, bipolar excitation.

one of the three schemes shown in Fig. 2.52. In the monofilar winding in (a), the magnetic polarity on excitation is always north or south, which implies that the polarity cannot be switched. This excitation method is termed unipolar excitation. In circuit (b), the direction of current in the coil can be switched because of a bridge circuit. However, four transistors are required for each phase. This method is referred to as the bipolar excitation. Circuit (c) involves a pair of bifilar windings and two transistors, by which the stator pole can be excited in any magnetic polarity, for one coil is used to excite a north pole while the other excites a south pole. Two coils wound in bifilar scheme are magnetically coupled when either is excited. If two independent coils are provided instead of bifilar windings, inductance differences will appear between the two coils and positioning accuracy will be degraded.

In general, the efficiency of a permanent-magnet motor operated in the alternating polarity mode is higher than the efficiency attained in the unipolar drive mode.

An advantage of bifilar windings in a single-stack VR motor is discussed in Section 2.3.5.

2.3 Modes of excitation

In the argument developed so far the principle of the stepping motor was explained in terms of the single-phase excitation. This excitation method is the most basic one and often used for analysing fundamental theoretical problems. However several different methods of excitation are in use today.

2.3.1 Single-phase excitation

Table 2.3 shows the sequences of a single-phase excitation mode for three- and four-phase motors. The shaded parts in the table represent the excited state, and the white blanks show the phases to which current is not supplied and so are not excited. When a motor revolves clockwise in the excitation sequence of Ph1 → Ph2 → Ph3 · · ·, it will revolve counterclockwise by simply reversing the sequence to Ph3 → Ph2 → Ph1 · · ·. Operation by single-phase excitation is also known as 'one-phase-on drive'.

2.3.2 Two-phase excitation operation

The operation of a motor in which two phases are always excited is called 'two-phase-on operation'. Before discussing the advantages of this method let us see the sequence of excitation and the relation between the rotor and stator teeth in the equilibrium position. The sequences are given in Table 2.4 for three- and four-phase motors. It is seen in these

Table 2.3. Excitation sequence in the single-phase-on operation.

(1) Three-phase motor

	R	1	2	3	4	5	6	7	8
Phase 1	▓			▓			▓		
Phase 2		▓			▓			▓	
Phase 3			▓			▓			▓

(2) Four-phase motor

	R	1	2	3	4	5	6	7	8
Phase 1	▓				▓				▓
Phase 2		▓				▓			
Phase 3			▓				▓		
Phase 4				▓				▓	
Pulses									

Note: Symbol R indicates 'reset'.

Table 2.4. Excitation sequence in the two-phase-on operation.

(1) Three-phase motor

Clock state	R	1	2	3	4	5	6	7	8
Phase 1	▨	▨	→	▨		▨	▨		
Phase 2		▨	→	▨	▨		▨		▨
Phase 3	▨		▨	▨	▨		▨	▨	▨

(2) Four-phase motor

Clock state	R	1	2	3	4	5	6	7	8
Phase 1	▨			▨	▨				▨
Phase 2		▨	▨			▨	▨		
Phase 3			▨	▨			▨	▨	
Phase 4	▨			▨	▨			▨	▨

tables that when an excitation current is switched from one phase to another (e.g. as shown by the arrow in Table 2.4(1) Ph2 is turned off and Ph1 is turned on) the third phase (Ph3 in the above example) remains excited.

The positional relation between the rotor and stator teeth in an equilibrium state is such as shown in Fig. 2.53. The teeth in both members are not in alignment as in the case of the single-phase-on mode shown in Fig. 2.13 on p. 28. The field distribution and step operation for

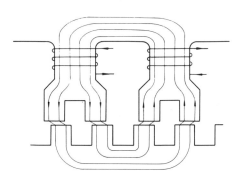

Fig. 2.53. Positional relation of rotor and stator teeth in the two-phase excitation.

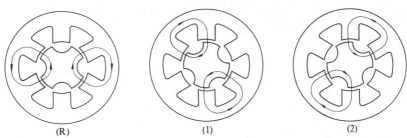

Fig. 2.54. Change in field pattern as a step proceeds in the two-phase-on mode in a monofilar three-phase VR motor.

the three-phase motor is illustrated in the cross-sectional view of Fig. 2.54.

A big characteristic difference between the single-phase-on and two-phase-on operations appears in the transient response as shown in Fig. 2.55. In the two-phase-on drive the oscillation damps more quickly than is the case of the single-phase-on mode. This can be explained qualitatively using Figs. 2.56 and 2.57 as follows. Two phases are always excited

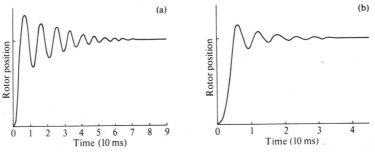

Fig. 2.55. Difference in single-step response between the single-phase (a) and two-phase (b) excitation.

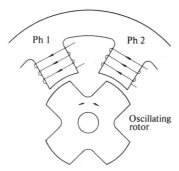

Fig. 2.56. Rotor oscillation in the two-phase excitation.

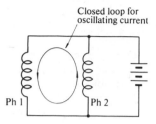

Fig. 2.57. Closed loop for the oscillating current.

in this mode of operation, and the circuit of the two phases forms a closed loop due to electromagnetic induction when oscillation occurs. This applies to the oscillating component of current, but not to the stationary component generating the holding torque. Thus, the oscillatory motion of the rotor results in oscillating current superimposed on the stationary current in each phase. It should be noticed that the phases of the oscillating component of current are opposite in Ph1 and Ph2. Since the torque generated by the oscillating component of the current acts in the opposite direction to the oscillatory motion, the oscillation is damped out. Or it may be considered that the kinetic energy associated with the rotor oscillation is converted into Joule heat through this process.

Since this type of closed circuit is not formed in the single-phase excitation mode, the oscillation is damped only by mechanical friction. This problem is discussed in detail in Section 4.3.

2.3.3 Half-step mode

The excitation scheme that is a combination of the single-phase and two-phase excitation is the so-called half-step operation. The excitation sequence for three-phase motors is given in Table 2.5. The numbers of clock states are here taken in two ways, (A) and (B). In the (A) way, positionings are made in the single-phase excitation only, and two phases are excited while moving from one equilibrium point to another. The

Table 2.5. Excitation sequence in the half-step operation (three-phase motor).

Clock state (A)	R	1		2		3		4		5	
Clock state (B)	R	1	2	3	4	5	6	7	8	9	
Phase 1		▨	▨				▨	▨	▨		
Phase 2			▨	▨	▨				▨	▨	
Phase 3				▨	▨	▨				▨	▨

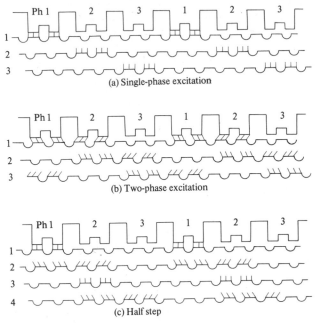

(a) Single-phase excitation

(b) Two-phase excitation

(c) Half step

Fig. 2.58. Comparison of the single-phase-on, two-phase-on, and half-step drives.

two-phase excitation is used here to suppress oscillation. In another method, the equilibrium positions of both the single and two-phase excitations are used for positioning. The clock states must be counted as in (B) in this mode. This scheme reduces the step angle to half.

Comparisons between the single-phase, two-phase, and half-step excitation in step operation for a three-phase motor are given in Fig. 2.58.

In most motors with more than four phases the half-step drive is carried out by a combination of two-phase and three-phase excitation or three- and four-phase excitation.

2.3.4   *Two-phase-on drive of bifilar-wound three-phase VR motor*

One of the most important requirements in motor design is to make the machine's size as small as possible for the demanded performance specifications. Pawletko and Chai[3] claim that a bifilar-wound three-phase VR motor operated in two-phase excitation meets this requirement. The schematic diagram of the winding connections is given in Fig. 2.59. It should be noted that the coils of the opposing poles are connected so that fluxes in both poles are directed either outward or inward at the same time. (Cf. Fig. 2.13 in which the two opposing poles generate different magnetic poles to each other.) Figure 2.60 shows the flux distribution resulting from the excitation sequence in Table 2.6. There are four flux

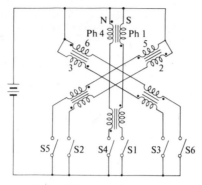

Fig. 2.59. Schematic diagram of the bifilar-wound coils in the three-phase VR motor.

loops distributed uniformly in the cores. This is in contrast to the excitation scheme in a monofilar-wound motor which has only two flux loops in both the one-phase-on and two-phase-on modes as illustrated in Figs. 2.13 and 2.54. Actually a bifilar-wound three-phase VR motor is much better in the torque-to-machine volume ratio and damping in comparison with corresponding monofilar wound motors.

2.3.5 *Excitation by a bridge circuit*

In the unipolar drive of a bifilar-wound hybrid motor, the windings are not fully utilized; e.g. only 50 per cent of the windings is used to carry

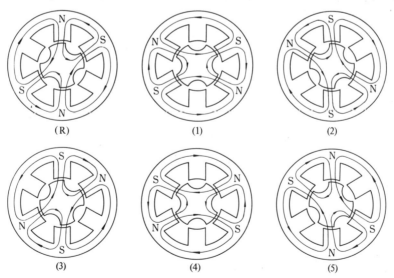

Fig. 2.60. Field pattern in the bifilar-wound three-phase motor. Symbol (R) denotes Reset in the table in Fig. 5.31 (p. 141).

Table 2.6. Excitation sequence in the two-phase-on operation for a bifilar-wound three-phase VR motor.

Clock state	R	1	2	3	4	5	6	7	8
Ph1 (S1)	■	■					■	■	
2 (S2)		■	■					■	■
3 (S3)			■	■					■
4 (S4)				■	■				
5 (S5)					■	■			
6 (S6)	■					■	■		

current at any time in the two-phase-on mode. Complete utilization of the windings by means of the bipolar drive considerably increases motor output power. The hybrid motor wound in the monofilar scheme may be regarded as a two-phase motor as shown in Fig. 2.61. Poles 1, 3, 5, and 7 belong to phase A (= PhA), and poles 2, 4, 6, and 8 to PhB in this model. For PhA if poles 1 and 5 generate magnetic north poles, poles 3 and 7

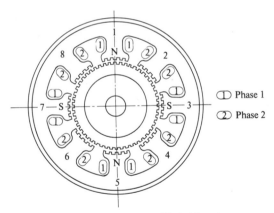

Fig. 2.61. Coil arrangement in a monofilar-wound hybrid motor.

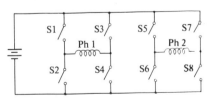

Fig. 2.62. Bridge driver scheme for a two-phase stepping motor.

Table 2.7. Excitation sequence in the bridge operation for a two-phase motor.

Clock state	R	1	2	3	4	5	6	7	8
S1	▨	▨			▨	▨			▨
S2			▨	▨			▨	▨	
S3				▨	▨			▨	▨
S4	▨	▨			▨	▨			
S5		▨	▨			▨	▨		
S6	▨	▨			▨			▨	▨
S7	▨			▨	▨			▨	▨
S8		▨	▨			▨	▨		

will produce magnetic south poles, and vice versa. A similar statement applies to PhB.

The bridge circuit shown in Fig. 2.62 is used to drive this motor in the bipolar mode, following the switching sequence given in Table 2.7. It is seen from these figures and tables that each winding always carries current alternating magnetic polarities. It is known that 25 to 30 per cent improvements in torque per unit of input power are possible by use of the bipolar drive. The only drawback of the bipolar drive is that it needs twice as many transistors as in the monopolar drive of a bifilar motor. If a centre-tapped supply is available, however, the number of switching devices is reduced to four as shown in Fig. 2.63. When a bifilar-wound

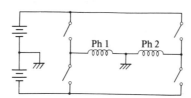

Fig. 2.63. Bridge driver scheme with a centre-tapped DC supply.

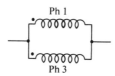

Fig. 2.64. Wire connection of a bifilar-wound hybrid motor when used in the bipolar drive.

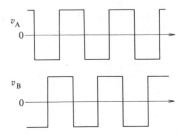

Fig. 2.65. Square-wave voltage for the bipolar drive.

motor is to be driven in the bipolar scheme, the wires should be connected as shown in Fig. 2.64 so that the windings of Ph1 and Ph3 may create the same sign of magnetic pole in each stator pole. The voltage waveform applied to each phase in the bipolar drive is a square wave as shown in Fig. 2.65.

2.3.6 *Ministep drive*

It is possible to subdivide one natural step into many small steps by means of electronics. This method is known as the ministep or microstep drive and is often applied to hybrid stepping motors. The idea of the ministep comes from the sinusoidal bipolar drive of a hybrid stepping motor as a synchronous motor. If a hybrid stepping motor is driven from a two-phase sine wave supply, instead of square wave, it is expected that the rotor motion is stepless and very smooth. This is true in some motors in a particular condition.[13] But in many cases a perfect stepless smooth motion is not realized due to detenting effect, variable reluctance effect and subharmonics induced in the voltage by the magnet. To subdivide one natural step, the supply current is shaped into a step-wave as in Fig. 2.66. If one cycle is divided into $4n$ sections, one natural step is resolved into n substeps. In an automated drafting system, which will be discussed in Section 8.2.3, the linear hybrid motor shown in Fig. 2.50 is operated by a 96-stepped sine wave excitation, to drive a draft head realizing a

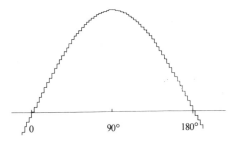

Fig. 2.66. A step-wave current for the ministep drive.

minimum 10 μm subdivision from a 0.96 mm tooth pitch. The line quality drawn on this system is excellent. When subdivisions are not so precise and motion is not very smooth, such as in the application of rotary hybrid stepping motors to graph plotters, an appropriate third harmonic component is superimposed on the current to eliminate adverse effects due to subharmonics in the induced voltage and other causes.[14],[15]

2.4 Single-phase stepping motors

All the motors described so far are polyphase stepping motors. There are, however, some stepping motors which are designed to be operated from a single-phase supply. They are widely used in watches and clocks, timers and counters. Current single-phase stepping motors all use one or two permanent magnets, because permanent magnets are quite necessary to raise the ratio of torque to input power in a miniature motor. In discussing single-phase stepping motors, we must consider two matters: (1) how to detent the rotor at a fixed position when the coil is not excited; and (2) how to rotate the rotor in the desired direction by switching the magnetic polarity of only one coil. We shall here consider these problems in respect to two typical motors.

2.4.1 *Single-phase stepping motor with asymmetrical air-gaps*

The motor illustrated in Fig. 2.67 has a cylindrically shaped magnet as its rotor while the air-gaps become narrower in one direction. The rotor will come to rest or detent either as in Figs. 2.68(a) or as in (b). The magnetically stable positions are such that the magnetic poles of the rotor come to the narrowest part of air-gaps. To cause the rotor detenting in the (a) position to rotate through 180°, the coil must be excited to produce flux in the direction shown in (a), since the magnetic polarities of the electromagnet and those of the permanent magnet repel each other in this state. As is obvious in this figure the natural direction of rotation is clockwise (CW) because of the unique air-gaps. If the coil is excited in the

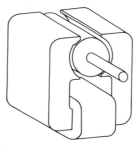

Fig. 2.67. A single-phase stepping motor.

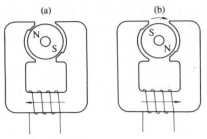

Fig. 2.68. Detent positions and coil polarity to rotate rotor.

opposite direction to that shown in (a), the rotor will be driven counter-clockwise through a little angle by an attractive torque between the permanent magnet and electromagnet. When excitation ceases, however, the rotor will move back to the previous detent position. To drive the rotor from position (b) to (a), the excitation must be given as indicated in (b). The rotational direction in this case is clockwise, too.

A stepping motor used in a wrist-watch employing this principle is shown in Fig. 2.69. The rotor is a rare-earth magnet disc of about 1.5 mm diameter. The stator core is not cut at both ends in this sample. As the two narrow parts are magnetically saturated when the coil is excited, the major flux from the excited coil will pass through the rotor. The applied potential to the coil is such as shown in Fig. 2.70; pulse width is as short as eight milliseconds to save the electrical energy of a small battery incorporated in the watch. In order to excite the correct polarity so as not

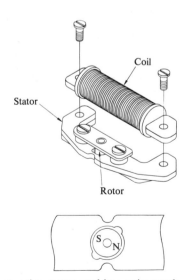

Fig. 2.69. A single-phase stepping motor used in a wrist-watch.

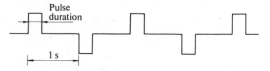

Fig. 2.70. Voltage waveform applied to a watch stepping motor.

to miss the first step when one starts a watch after synchronizing it with a standard time, a circuit to memorize rotor position and excite in the correct polarity is incorporated.

2.4.2   Cyclonome®

A stepping motor manufactured by Sigma Instruments, Inc., and called Cyclonome®, is interesting. Its basic structure is illustrated in Fig. 2.71; two L-shaped Alnico V magnets of high coercive force are located in the stator, while the rotor is of soft iron having a number of teeth, 10 in this example. The stator has three poles X, Y, and Z made of laminated soft steel. Poles X and Y, called the driving poles, have three teeth with a tooth pitch of 36°, and pole Z is the detent pole. Let us study the principle of this single-phase motor[16] with reference to Fig. 2.72. State (1) is a detent position or no-current stable position. The teeth 2, 4, and 6 in the detent pole Z are aligned with three of the rotor teeth to pass the magnet flux to the rotor. Since the gaps at pole Y are small in this state, most of the flux passes through from rotor to pole Y. State (2) illustrates the equilibrium position when a positive current is applied to the coil. The flux at driving pole Y is neutralized, while flux is increased at pole X, which causes a CW movement and results in alignment of teeth at the pole X. State (3) represents the second detent position, secured by alignment of detent teeth 1, 3, and 5 with three rotor teeth. The rotor has rotated 18° from the state (1) position. State (3) shows the rotor position when the coil is supplied by a minus current; the teeth at pole Y are brought into alignment with three of the rotor teeth by reinforced flux at

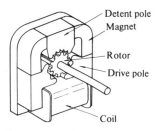

Fig. 2.71. Structure of the Cyclonome stepping motor.

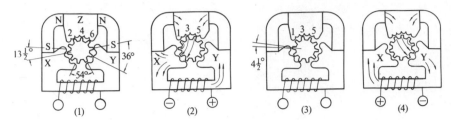

Fig. 2.72. Principle of the Cyclonome stepping motor.

that pole. The position (1) will be restored when the excitation is again turned off. The rotor has revolved clockwise through 36°.

2.5   Specification of stepping motor characteristics

In this section, technical terms used for specifying the characteristics of a stepping motor are studied.

2.5.1   *Static characteristics*

The characteristics relating to stationary motors are called static characteristics.

(1) *T/θ characteristics.* The stepping motor is first kept stationary at a rest (equilibrium) position by supplying a current in a specified mode of excitation, say, single-phase or two-phase excitation. If an external torque is applied to the shaft, an angular displacement will occur. The relation between the external torque and the displacement may be plotted as in Fig. 2.73. This curve is conventionally called the T/θ characteristic curve, and the maximum of static torque is termed the 'holding torque', which

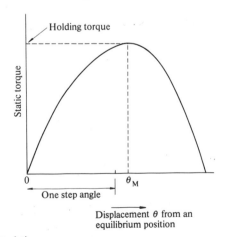

Fig. 2.73. T/θ characteristics.

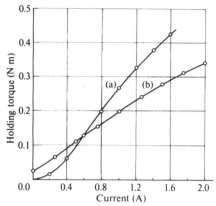

Fig. 2.74. Examples of T/I characteristics: (a) a 1.8° four-phase VR motor; and (b) a 1.8° four-phase hybrid motor. (After Ref. [17].)

occurs at $\theta = \theta_M$ in Fig. 2.73. At displacements larger than θ_M, the static torque does not act in a direction towards the original equilibrium position, but in the opposing direction towards the next equilibrium position. The holding torque is rigorously defined as 'the maximum static torque that can be applied to the shaft of an excited motor without causing continuous motion'. The angle at which the holding torque is produced is not always separated from the equilibrium point by one step angle.

(2) *T/I characteristics.* The holding torque increases with current, and this relation is conventionally referred to as *T/I* characteristics. Figure 2.74[17] compares the *T/I* characteristics of a typical hybrid motor with those of a VR motor, the step angle of both being 1.8°. The maximum static torque appearing in the hybrid motor with no current is the detent torque as defined in Section 2.1.3.

For details of the measurement of torques and displacements see Reference [18].

2.5.2 Dynamic characteristics

The characteristics relating to motors which are in motion or about to start are called dynamic characteristics.

(1) *Pull-in torque characteristics.* These are alternatively called the starting characteristics and refer to the range of frictional load torque at which the motor can start and stop without losing steps for various frequencies in a pulse train. The number of pulses in the pulse train used for the test is 100 or so. The reason why the word 'range' is used here, instead of 'maximum', is that the motor is not capable of starting or maintaining a normal rotation at small frictional loads in certain frequency ranges as

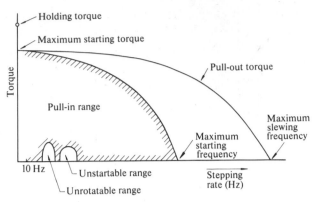

Fig. 2.75. Dynamic characteristics.

indicated in Fig. 2.75. When the pull-in torque is measured or discussed, it is also necessary to specify clearly the driving circuit, the measuring method, the coupling method, and the inertia to be coupled to the shaft. In general, the self-starting range decreases with increases in inertia.

(2) *Pull-out torque characteristics.* This is alternatively called the slewing characteristic. After the test motor is started by a specified driver in the specified excitation mode in the self-starting range, the pulse frequency is gradually increased; the motor will eventually run out of synchronism. The relation between the frictional load torque and the maximum pulse frequency with which the motor can synchronize is called the pull-out characteristic (see Fig. 2.75). The pull-out curve is greatly affected by the driver circuit, coupling, measuring instruments, and other conditions.

(3) *The maximum starting frequency.* This is defined as the maximum control frequency at which the unloaded motor can start and stop without losing steps.

(4) *Maximum pull-out rate.* This is defined as the maximum frequency (stepping rate) at which the unloaded motor can run without losing steps, and is alternatively called the 'maximum slewing frequency'.

(5) *Maximum starting torque.* This is alternatively called 'maximum pull-in torque' and is defined as the maximum frictional load torque with which the motor can start and synchronize with the pulse train of a frequency as low as 10 Hz.

References for Chapter 2

[1] Biscoe, G. I. and Mills, A. S. (1977). The rationalisation and standardization of stepping motors and their test methods. *Proc. Sixth Annual Symposium on*

Incremental motion control systems and devices. Department of Electrical Engineering, University of Illinois. pp. 331–42.

[2] Kenjo, T. and Niimura, Y. (1979). *Fundamentals and applications of stepping motors*. (In Japanese.) pp. 243–4. Sogo Electronics Publishing Co., Ltd., Tokyo.

[3] Pawletko, J. P. and Chai, H. D. (1976). Three-phase variable-reluctance step motor with bifilar winding. *Proc. Fifth Annual Symposium on Incremental motion control systems and devices*. Department of Electrical Engineering, University of Illinois. pp. F1–F8.

[4] Egawa, K. (1971). Stator design for a stepping motor. United States Patent No. 3,601,640.

[5] Feiertag, K. M. and Donahoo, J. T. (1952). Dynamoelectric machine. United States Patent No. 2,589,999.

[6] Heine, G. (1977). Control methodology of 5-phase PM stepping motors. *Proc. Sixth Annual Symposium on Increment motion control systems and devices*. Department of Electrical Engineering, University of Illinois. pp. 313–30.

[7] Kuo, B. C. (1979). *Step motors and control systems*. Chapter 14. SRL Publishing Company, Champaign, Illinois.

[8] Goddijn, B. H. A. (1978). Effect of the number of pole pairs and permanent magnet excitation on the performance of a hybrid stepping motor. *Proc. Seventh Annual Symposium on Incremental motion control systems and devices*. Incremental Motion Control Systems Society, Champaign, Illinois. pp. 7–12.

[9] Niimura, Y. (1974). Outer-rotor-type stepping motor. *Proc. Third Annual Symposium on Incremental motor control systems and devices*. Department of Electrical Engineering, University of Illinois. pp. H1–H10.

[10] Pawletko, J. P. (1976). Dynamic responses and control aspects of linear stepping motors. *Proc. Fifth Annual Symposium on Incremental motor control systems and devices*. Department of Electrical Engineering, University of Illinois. pp. P1–P17.

[11] Hinds, W. E. and Nocito, B. (1974). The Sawyer linear motor. *Proc. Third Annual Symposium on Incremental motor control systems and devices*. Department of Electrical Engineering, University of Illinois. pp. W1–W10.

[12] Langley, L. W. and Kidd, H. K. (1979). Closed-loop operation of a linear stepping motor under microprocessor control. *Proc. International conference on stepping motors and systems*. University of Leeds, England. pp. 32–6.

[13] Kenjo, T. and Takahashi, H. (1979). Speed ripple characteristics of hybrid stepping motors driven in the ministep mode. *Proc. International conference on stepping motors and systems*. University of Leeds, England. pp. 87–93.

[14] Patterson, M. L. and Haselby, R. D. (1977). A microstepped XY controller with adjustable phase current waveforms. *Proc. Sixth Annual Symposium on Incremental motor control systems and devices*. Department of Electrical Engineering, University of Illinois. pp. 163–8.

[15] Patterson, M. L. (1977). Analysis and correction of torque harmonics in permanent-magnet step motors. Ibid. pp. 25–37.

[16] Bailey, S. J. (1960). Incremental servos, Part II: Operation and analysis. *Control Engineering* Dec. pp. 97–102.

[17] Kenjo, T. and Niimura, Y. (1979). *Fundamentals and applications of stepping motors*. (In Japanese.) p. 231. Sogo Electronics Publishing Co., Ltd., Tokyo.

[18] Kuo, B. C. (1979). *Step motors and control systems*. Chapter 6. SRL Publishing Company, Champaign, Illinois.

3. Theory of electromagnetics and structure of stepping motors

In Chapter 2 a qualitative approach was employed for explaining how torque is produced in a stepping motor; it was an explanation in terms of the tension in lines of magnetic force. In this chapter the mechanism of torque production will be analysed quantitatively using an electrodynamic approach. This is followed by discussions on tooth-structure problems in stepping motors.

The main symbols used in this and the next chapter are shown in Table 3.1.

3.1 Mechanism of static torque production in a VR stepping motor

There are several ways of expressing the torque developed in an electrical motor. The qualitative explanation given in Chapter 2 may be followed by a theory in terms of the Maxwell stress tensor. But this approach, which is basically field theory, is not always suitable for the treatment of stepping motors in terms of circuitry parameters. In this chapter, instead, a theory in terms of magnetic energy and coenergy is presented. We will start with the ideal case in which the rotor and stator cores have infinite permeability, and proceed step by step to the case in which the cores are subject to magnetic saturation.

3.1.1 *The case of infinitely permeable cores*

To analyse the situation of an iron piece being drawn into a magnetic field created by an electromagnet as shown in Fig. 3.1, we use the model of Fig. 3.2. A current I is flowing through the coil of n turns to yield magnetic flux, and a force f is acting on the iron piece in the x-direction. The iron piece may be regarded as a tooth of the rotor of a stepping motor, and the electromagnet corresponds to a pair of teeth of the stator in a VR motor. First let us determine the magnetic flux density B_g in the air-gaps (which are the spaces indicated by $g/2$ in the figure). Ampere's circuital law along the dotted closed loop is expressed as

$$\oint \mathbf{H} \cdot d\mathbf{l} = nI. \tag{3.1}$$

The left-hand side of this equation is rewritten as

$$\oint \mathbf{H} \cdot d\mathbf{l} = H_g\left(\frac{g}{2}\right) + H_g\left(\frac{g}{2}\right) + H_i l = H_g g + H_i l, \tag{3.2}$$

Table 3.1. Main symbols used in Chapters 3 and 4.

B	Magnetic flux density: B_g flux density in gap; B_s saturation level of flux density (T)
$\cdot C$	Constant determined from motor dimensions and number of turns (V s rad^{-1}, N m A^{-1} rad^{-1})
D	Viscous friction coefficient (N m s rad^{-1})
d	Tooth depth (m)
E_m	$K_T I_M N_r$, see eqn (4.92)
e	Electromotive force (V)
f	Force (N)
g	Gap length (m)
H	Magnetic field intensity: H_g field intensity in gap (A m^{-1})
i, I	Current (A)
J	Moment of inertia (kg m^2) Non-dimensional inertia-ratio, see eqn (4.99)
k_p, k_v	Motor constants, see eqns (4.23) and (4.39), respectively
K_T	Torque constant (N m A^{-1} rad^{-1})
L	Self inductance (H)
m	Number of phases
M	Mutual inductance (H)
n	Number of turns
N_r	Number of rotor teeth
N_s	Number of stator teeth
p	Number of pole pairs
q	Number of stator teeth per phase
r, R	Resistances (Ω)
s	Laplace operator d/dt (s^{-1})
S	Number of steps per revolution (rad^{-1})
T	Torque: T_p pull-out torque; T_M maximum static torque
t	Time
v, V	Voltage (V)
w	Tooth width (m)
W_m	Magnetic energy (J)
x	Displacement in the x-direction, overlapped length (m)
α	Time constant (s)
β	Time constant (s)
δ, Δ	Perturbation (non-dimensional)
ζ	Damping ratio (non-dimensional)
θ, Θ	Rotational angle (rad)
λ	Tooth pitch (rad)
μ	Permeability: μ_0 permeability in gap (H/m)
ξ	Rotational angle in terms of electrical angle (rad)
ρ	Torque angle (rad)
τ	Torque (N m)
ϕ, Φ	Magnetic flux (T m^2)
ψ, Ψ	Flux linkages (T m^2)
ω, Ω	Angular speed (rad s^{-1})
ω_n	Natural frequency (s^{-1})
$\cdot$	d/dt (s^{-1})

Note (1) Variables in lower case letters are functions of time.
     (2) Variables as a function of s are designated by capital letters.
     (3) τ and T are used for torque.

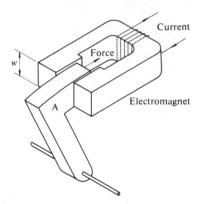

Fig. 3.1. An iron piece is attracted by an electromagnet.

where H_g = magnetic field intensity in the gaps,
         H_i = magnetic field intensity in the cores,
          l = total magnetic path in the cores.
When the permeability of cores is extremely large, H_i is so low that it is allowable to put $H_i = 0$. If $H_i = 0$ and the core permeability μ is infinity, a physical absurdity results, namely that $B = \mu H_i = \infty$ in the cores. H_g is hence given by

$$H_g = nI/g. \tag{3.3}$$

The gap flux density is

$$B_g = \mu_0 nI/g, \tag{3.4}$$

where μ_0 is the permeability in the gap length.

Let the transverse length of the iron piece be w, and let the distance by which the rotor tooth and the iron piece overlap be x (see Fig. 3.3). The overlapped area is now xw. The B_g in eqn (3.4) multiplied by the overlapped area is the magnetic flux:

$$\Phi = xw\mu_0 nI/g. \tag{3.5}$$

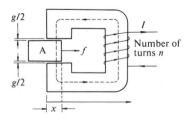

Fig. 3.2. A model for a stepping motor.

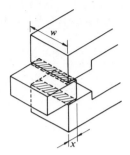

Fig. 3.3. Overlapped area.

Hence the flux linkages Ψ is given by

$$\Psi = n\Phi = xw\mu_0 n^2 I/g. \tag{3.6}$$

Now let us assume that there is an incremental displacement, Δx, of the tooth during a time interval Δt as illustrated in Fig. 3.4. Then the increment in the flux linkage, $\Delta\Psi$, is

$$\Delta\Psi = \frac{w\mu_0 n^2 I}{g} \Delta x. \tag{3.7}$$

The e.m.f. induced in the coils by the change in flux linkage is

$$e = -\frac{\Delta\Psi}{\Delta t} = -\frac{w\mu_0 n^2 I}{g}\frac{\Delta x}{\Delta t}. \tag{3.8}$$

The minus sign in this equation implies that the direction of the e.m.f. is opposing the current. Since the current I is supplied by the power source for the time interval Δt overcoming the counter-e.m.f., the work ΔP_i done by the source is

$$\Delta P_i = I\,|e|\,\Delta t = \frac{w\mu_0 n^2 I^2}{g}\Delta x. \tag{3.9}$$

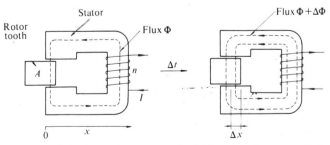

Fig. 3.4. A rotor tooth is drawn by a magnetic field and undergoes a displacement Δx during a time interval Δt.

The coil resistance is here assumed to be zero to simplify analysis. Using eqn (3.4), ΔP_i is expreseed in terms of B_g as follows:

$$\Delta P_i = \frac{B_g}{\mu_0} gw \, \Delta x. \tag{3.10}$$

The work done by the source is converted partly to mechanical work, and the rest is spent increasing the magnetic field energy in the gaps. The increase in the gap field energy is given by

$$\Delta W_m = \frac{1}{2} \frac{B_g^2}{\mu_0} \times (\text{the increase in the gap space})$$

$$= \frac{1}{2} \frac{B_g^2}{\mu_0} gw \, \Delta x. \tag{3.11}$$

From observation of eqns (3.10) and (3.11) we can find that a half of ΔP_i is converted into the magnetic field energy in the gaps. Consequently we are allowed to say that the other half of ΔP_i is converted into the mechanical work. Since the mechanical work is the force f multiplied by the displacement Δx, we obtain

$$f\Delta x = \frac{1}{2} \frac{B_g^2}{\mu_0} gw \, \Delta x. \tag{3.12}$$

Eliminating Δx from both sides

$$f = \frac{1}{2} \frac{B_g^2}{\mu_0} gw, \tag{3.13}$$

which, by use of eqn (3.4), may be put in the form

$$f = \frac{1}{2} \frac{w\mu_0 n^2 I^2}{g}. \tag{3.14}$$

On the other hand, the magnetic energy W_m in the gap is

$$W_m = \frac{1}{2} \frac{B_g^2}{\mu_0} gxw. \tag{3.15}$$

From eqns (3.13) and (3.15), therefore, we derive

$$f = \frac{dW_m}{dx}. \tag{3.16}$$

Attention must, however, be paid to the assumption that the current I is kept constant during the displacement. Hence eqn (3.16) must be described in the rigorous form

$$f = \left(\frac{\partial W_m}{\partial x}\right)_{I=\text{const}} \tag{3.17}$$

This equation is valid for the general case in which the coil resistance is not zero. On the other hand, if we employ a model in which the flux is kept constant during the displacement,[1] then we will obtain the form

$$f = -\left(\frac{\partial W_m}{\partial x}\right)_{\Phi = \text{const}}. \tag{3.18}$$

In dealing with stepping motors, eqn (3.17) is more useful than eqn (3.18).

3.1.2 *The case of constant permeabilities*

In the model with infinitely permeable cores, the magnetic field appears only in the gaps, and its mathematical treatment is simple. When cores are of finite permeability, on the other hand, magnetic energy appears not only in the gaps, but also in the cores and spaces other than the gaps. It is not easy to analyse such situations by means of electromagnetic field theory. Instead we will derive an expression for force in terms of circuitry parameters under the assumption that the permeabilities are not the functions of magnetic field.

If the coil inductance is L in the model of Fig. 3.3, the flux linkages Ψ is given by

$$\Psi = LI. \tag{3.19}$$

The magnetic energy W_m in the system is expressed as

$$W_m = \tfrac{1}{2}LI^2. \tag{3.20}$$

If the iron piece undergoes a displacement Δx during the time interval Δt, the inductance L will increase by ΔL. The e.m.f. induced in the coil is

$$e = -\frac{\Delta\Psi}{\Delta t} = -\frac{\Delta(LI)}{\Delta t}. \tag{3.21}$$

If the power supply is a current source and provides a current I during the displacement, eqn (3.21) is simplified as

$$e = -I\frac{\Delta L}{\Delta t}. \tag{3.22}$$

Since the voltage at the source is equal but opposite to the counter-e.m.f. of eqn (3.22), the work ΔP_i done by the source on the circuit is

$$\Delta P_i = I\,|e|\,\Delta t = I^2\,\Delta L. \tag{3.23}$$

On the other hand, the increase in the magnetic energy ΔW_m is

$$\Delta W_m = \tfrac{1}{2}I^2\,\Delta L. \tag{3.24}$$

From comparison of eqns (3.23) and (3.24), it is seen that a half of the work done on the circuit by the source is converted into magnetic energy. Hence it is supposed that the other half is converted to mechanical work ΔP_0;

$$\Delta P_0 = f \, \Delta x = \tfrac{1}{2} I^2 \, \Delta L. \tag{3.25}$$

Then the force is

$$f = \tfrac{1}{2} I^2 \frac{\Delta L}{\Delta x}. \tag{3.26}$$

In the above procedure it was assumed that the coil resistance was zero and the power supply was a current source. But eqn (3.26) can be applied to general cases. Thus the force developed on the iron piece is in the direction which will increase the inductance or lessen the reluctance.

3.1.3 *Treatment of magnetic saturation*

In most stepping motors the cores are subject to magnetic saturation. If a motor is designed to be operated in the linear B/H characteristic region, the torque produced per unit volume will be so small that the motor is too big to serve in practical applications. For this reason, a theory which does not take account of any saturation is impracticable. A general theory for torque is developed here to deal with magnetic saturation in cores.

Again using the model of Fig. 3.3 let us analyse the energy conversion. The iron piece or tooth is drawn by a force f due to the magnetic field induced by the coil current, and travels from x_0 to $x_0 + \Delta x$ taking a time Δt. The flux interlinkage ψ is a function of the position x and the current i in this case, and expressed by $\psi(x, i)$. In the analysis here the variables are expressed by lower case letters, ψ, i. If the current i is kept at value I during the displacement, the work ΔP_i done by the power supply for the time interval Δt is

$$\Delta P_i = Iv \, \Delta t = I \frac{\Delta \Psi}{\Delta t} \Delta t = I \, \Delta \Psi. \tag{3.27}$$

On the other hand the mechanical work done on the iron piece during the interval Δt is

$$\Delta P_0 = f \, \Delta x. \tag{3.28}$$

The increase in the magnetic energy in the system during the displacement Δx is expressed by

$$\Delta W_m = \int_0^{\Psi + \Delta \Psi} i \, d\psi(x_0 + \Delta x, i) - \int_0^{\Psi} i \, d\psi(x_0, i). \tag{3.29}$$

In this equation the current i is treated as a variable which varies from 0 to I as ψ varies from 0 to the final value $\Psi + \Delta\Psi$ or Ψ. The physical interpretation for each term in the right-hand side is as follows:

First term: The magnetic energy of the system in which the iron piece is positioned at $x = x_0 + \Delta x$. The integration should be done with respect to ψ from 0 to $\Psi + \Delta\Psi$, with x fixed at $x_0 + \Delta x$. (See Fig. 3.5(b).)

Second term: The magnetic energy of the system in which the iron piece is positioned at $x = x_0$. The integration should be done with respect to ψ from 0 to Ψ, with x fixed at x_0. (See Fig. 3.5(a).)

Each term is integrated by parts as follows.
First term:

$$\int_0^{\Psi+\Delta\Psi} i \, \mathrm{d}\psi = [i\psi]_0^{I(\Psi+\Delta\Psi)} - \int_0^I \psi(x_0 + \Delta x, i) \, \mathrm{d}i$$

$$= I(\Psi + \Delta\Psi) - \int_0^I \psi(x_0 + \Delta x, i) \, \mathrm{d}i. \qquad (3.30)$$

Second term:

$$\int_0^{\Psi} i \, \mathrm{d}\psi = I\Psi - \int_0^I \psi(x_0, i) \, \mathrm{d}i. \qquad (3.31)$$

The second terms in the right-hand sides of these two equations are in the form of magnetic coenergy (see Fig. 3.5). Substituting these forms into eqn (3.29) we obtain

$$\Delta W_m = I \, \Delta\Psi - \left\{ \int_0^I \psi(x_0 + \Delta x, i) \, \mathrm{d}i - \int_0^I \psi(x_0, i) \, \mathrm{d}i \right\}. \qquad (3.32)$$

Since the second term is the change in the magnetic coenergy associated with the rotor displacement Δx, eqn (3.32) may be written as

$$\Delta W_m = I \, \Delta\Psi - \Delta \int_0^I \psi(x, i) \, \mathrm{d}i. \qquad (3.33)$$

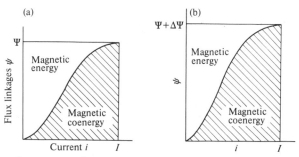

Fig. 3.5. Magnetic energy and coenergy at two different positions: (a) at $x = x_0$; (b) at $x = x_0 + \Delta x$.

Since the first term on the right-hand side is the work done by the power supply as indicated in eqn (3.27), we may rewrite eqn (3.33) as follows:

$$\Delta P_i = \Delta W_m + \Delta \int_0^I \psi(x, i)\, di. \qquad (3.34)$$

On the other hand we have

$$\Delta P_i = \Delta W_m + \Delta P_0. \qquad (3.35)$$

From comparison of eqns (3.34) and (3.35), we obtain, for the mechanical work ΔP_0, the relation

$$\Delta P_0 = f\, \Delta x = \Delta \int_0^I \psi(x, i)\, di, \qquad (3.36)$$

from which we obtain the expression for force:

$$f = \frac{\partial \int_0^I \psi(x, i)\, di}{\partial x} = \left(\frac{\partial\, (\text{magnetic coenergy})}{\partial x}\right)_{I=\text{const}}. \qquad (3.37)$$

The corresponding torque expression is

$$T = \frac{\partial \int_0^I \psi(\theta, i)\, di}{\partial \theta} = \left(\frac{\partial\, (\text{magnetic coenergy})}{\partial \theta}\right)_{I=\text{const}}, \qquad (3.38)$$

where θ is the angular position of the rotor.

These are the fundamental equations necessary to calculate the force and torque produced in stepping motors when magnetic saturation affects the machine characteristics. When a system has n coils the torque equation is expressed as

$$T = \frac{\partial}{\partial \theta} \sum_{j=1}^{n} \int_0^{I_j} \psi(\theta, i_j)\, di_j. \qquad (3.39)$$

3.1.4 *Effect of saturation to improve efficiency*

Let us discuss an effect of saturation using the model shown in Fig. 3.6. One of the features of this model is that the gap size is extremely small and it is allowed to put $g = 0$ for approximation in analysis. Another feature is that the iron piece has a rectangular B/H characteristic; the saturation level is denoted by B_s. It is assumed that the average permeabilities $(= B_s/H)$ are much higher than the gap permeability μ_0 but much lower than the stator core permeability:

$$\mu_0 \ll \frac{B_s}{H} \ll \text{permeability of stator core}. \qquad (3.40)$$

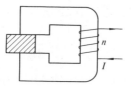

Fig. 3.6. A model for negligible gap and a tooth of rectangular saturable material.

Under these conditions the magnetic energy stored in the system is so small that the work done by the source is almost all converted into mechanical work, which is explained in the following.

In a space where the permeability μ is constant the magnetic energy stored per unit volume is $B^2/2\mu$. Firstly it is found that the magnetic energy in the stator core is low, since the permeability is assumed to be very large. Secondly, the gap volume is so small that the magnetic energy in this region is also low. The next problem is the estimation of the magnetic energy in the iron piece drawn by the magnetic field. The material of the iron piece is assumed to have a non-linear rectangular B/H characteristic. The magnetic energy in a unit volume is generally expressed by $\int_0^B H\, dB$, and the value of this integration is equal to the hatched areas in the graphs of Fig. 3.7. The curve of Fig. 3.7(a) represents the general B/H relation. For a linear B/H relation as illustrated in graph (b) the hatched area is obviously $(1/2)HB$ or $B^2/2\mu$. But for the saturable characteristic of graph (c) the hatched area is negligibly small, which means that little magnetic energy is stored in the space.

Now let us consider with the above preparation, how electric energy is converted into mechanical work. The flux ϕ flowing through the area in which the stator core and the iron piece are overlapping each other is given by

$$\phi = B_s xw. \tag{3.41}$$

The flux linkage ψ is

$$\psi = nB_s xw. \tag{3.42}$$

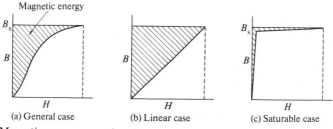

Fig. 3.7. Magnetic energy per unit volume equals the hatched area.

Hence the e.m.f. induced as the rotor moves is

$$-e = nB_s w \frac{dx}{dt}.$$ (3.43)

The work done by the current source per unit time, or the electric power, is

$$-eI = InB_s w \frac{dx}{dt}.$$ (3.44)

As seen previously, this work may be converted into magnetic energy and mechanical work. However, since the stored magnetic energy is negligible in the present case, the electric power will be converted mostly into mechanical output power. The mechanical work done on the iron piece is the force f multiplied by the speed dx/dt; consequently we obtain

$$f\frac{dx}{dt} = InB_s w \frac{dx}{dt}.$$ (3.45)

Therefore

$$f = InB_s w.$$ (3.46)

It should be noted here that when the B/H relation is linear and the final value of the flux density equals B_s, the force is expressed as

$$f = \tfrac{1}{2}InB_s w.$$ (3.47)

We can also deduce eqn (3.46) from the magnetic coenergy principle, noting that the coenergy per unit volume in the iron piece is $\int_0^B B\,dH = B_s H$ and the coenergy in the rest is negligible.

By comparison of eqns (3.46) and (3.47) it may be claimed that the torque developed in a stepping motor using a saturable steel for the rotor can be twice as high as that produced in a motor using material with a linear B/H characteristic. In order to realize this torque, however, the gap length must be as short as possible to minimize the magnetic energy stored there. When the gap length is not zero, the factor is less than two. Byrne[2] derived a similar conclusion in the analysis of the force produced between two partially overlapping teeth, and Lawrenson et al.[3] showed some experimental results which verify that the force on a strongly saturated iron piece in models similar to that of Fig. 3.6 is more than the force calculated for the linear B/H characteristic curves.

Moreover, the electrical loss in the conductor may be reduced by the use of saturable steel in the rotor, which is explained in the following. The coil resistance was assumed to be zero in the preceding analyses to simplify the derivation of expressions for torque and force. When we discuss the Joule loss, however, winding resistance must be taken into consideration. In a stepping motor system under normal operation, the magnetic energy in each phase of the windings goes and comes between the motor and power supply. In some drives, magnetic energy is dissipated into conduction loss during the main transistor's off period when the exciting current collapses through the suppressor circuit. In some circuits, the magnetic energy returns to the power supply through feedback diodes (see Figs. 4.28 and 4.30); but some of the magnetic energy is dissipated as Joule heating in the winding resistance and other parts of the circuit. Therefore, if no magnetic energy is stored in the system, useless losses are reduced and the motor efficiency will be improved.

The effect of saturation described above is valid for the model of Fig. 3.6. A practical machine based on the principles of this model has the rotor teeth revolving in a space sandwiched by the stator teeth. Stepping motors of this design were once manufactured. Nowadays this design is not favoured because of difficulties in assembly. But it is known that even when the materials used for the rotor and stator cores are ordinary silicon steel, this design produces good static torque.

3.1.5 Torque versus displacement characteristic in doubly salient structure

Most stepping motors nowadays have a doubly salient structure for cores. A typical tooth design is shown in Fig. 3.8; this is employed for a stack of a multi-stack VR motor. The torque T vs. displacement θ characteristics (T/θ characteristics) are shown in Fig. 3.9.[4] The curve at 0.16 of the rated current is quite flat, which agrees fairly well with the theoretical result expressed by eqn (3.14), which was derived on the assumption that the teeth have infinite permeability. At the rated current the shape is very different, approximating more nearly to a sine wave. The increase of the maximum torque when the current is increased from 0.5 to 1.33 of the rated current is by no means proportional to I^2, which would be the case if the flux current relationship were linear (see Fig. 3.10).

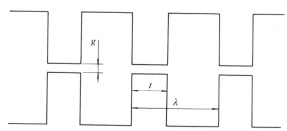

Fig. 3.8. Toothed structure employed for a multi-stack VR motor, drawn in linear motor model.

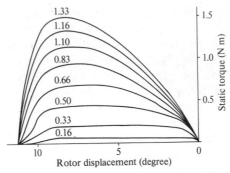

Fig. 3.9. T/θ characteristic curves; alignment at $0°$; curves are labelled with corresponding currents expressed per unit of rated value. (after Ref. [4], reproduced by permission of the Institution of Electrical Engineering, and by courtesy of Professor P. J. Lawrenson, Professor M. R. Harris and Dr A. Hughes).

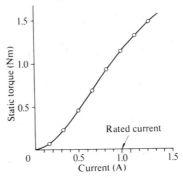

Fig. 3.10. Maximum static torque versus current.

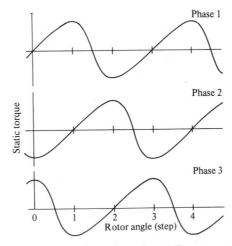

Fig. 3.11. An example of T/θ curves for a three-phase VR motor.

The T/θ characteristics over more than one step angle with a single phase excited in a three-phase motor are such as shown in Fig. 3.11. The points at which the curve crosses the horizontal axis where it has a negative slope are the rest or equilibrium points. This kind of curve is approximated by a sinusoidal wave or a straight line in analyses, depending on the problem under discussion.

3.1.6 *Effects of mutual induction*

As stated in Chapter 2, there are a number of drive schemes for stepping motors: one-phase-on drive, two-phase-on drive, three-phase-on drive, half-step drive, etc. In a drive scheme other than the one-phase-on drive, it is desirable that the mutual inductance is a minimum, since the mutual inductance has a general tendency to degrade the accuracy in positioning. When mutual inductance is not negligible, the torque in terms of linear theory is derived by the following procedure.

Let us discuss the case depicted in Fig. 3.12. When the power sources are current sources, induced voltages at each phase are

$$e_1 = -I_1 \frac{\Delta L_1}{\Delta t} - I_2 \frac{\Delta M}{\Delta t} \qquad (3.48)$$

$$e_2 = -I_2 \frac{\Delta L_2}{\Delta t} - I_1 \frac{\Delta M}{\Delta t} \qquad (3.49)$$

where L_1 = inductance of phase 1
L_2 = inductance of phase 2
M = mutual inductance between the two phases.
The work ΔP_i done by the two power supplies during the time increment Δt is

$$\begin{aligned}\Delta P_i &= -(e_1 I_1 + e_2 I_2)\,\Delta t \\ &= I_1^2\,\Delta L_1 + I_2^2\,\Delta L_2 + 2I_1 I_2\,\Delta M.\end{aligned} \qquad (3.50)$$

On the other hand the increment of the magnetic energy in the system is

$$\Delta W_m = \tfrac{1}{2}(I_1^2\,\Delta L_1 + I_2^2\,\Delta L_2) + I_1 I_2\,\Delta M. \qquad (3.51)$$

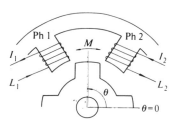

Fig. 3.12. A model for the case when mutual inductance does occur.

From eqns (3.50) and (3.51), it is seen that a half of the work done by the sources is converted to an increment of the magnetic energy. Consequently the other half is converted into the mechanical output $\Delta P_0 = T \Delta \theta$:

$$T \Delta \theta = \tfrac{1}{2}(I_1^2 \Delta L_1 + I_2^2 \Delta L_2) + I_1 I_2 \Delta M.$$

Hence the torque is

$$T = \tfrac{1}{2}I_1^2 \frac{\partial L_1}{\partial \theta} + \tfrac{1}{2}I_2^2 \frac{\partial L_2}{\partial \theta} + I_1 I_2 \frac{\partial M}{\partial \theta}. \qquad (3.52)$$

3.2 Theory of torque produced in hybrid stepping motors

The torque produced in a hybrid stepping motor is discussed here. One of the differences from the theory for the VR motor is that the action of the permanent magnet is to be taken into account. Patterson[5] presented a torque theory for hybrid motors including torque ripples. But only the stationary torque will be discussed here.

3.2.1 *Analytical approach*

We deal with a two-phase motor having the pole configuration shown in Fig. 3.13. (The teeth on the rotor are not drawn.) The stator windings are interconnected as shown to form a bipolar two-phase arrangement; the coils on poles 1, 3, 5, 7 are connected in series to compose phase A, and the coils on poles 2, 4, 6, 8 are in series to compose phase B. To simplify

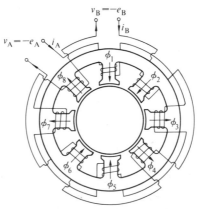

Fig. 3.13. A model of a hybrid stepping motor with windings connected in a two-phase bifilar arrangement.

the analysis, the effects of winding resistance, eddy currents, detent torque, mutual induction, and hysteresis are neglected. Also magnetic circuits in the motor are assumed to be linear, that is the magnetic flux induced by stator currents is independent of the internal magnet and proportional to the applied e.m.f.

Let us start from the fundamental law of energy conservation

$$\begin{pmatrix}\text{Electric power}\\ \text{supplied by}\\ \text{source}\end{pmatrix} = \begin{pmatrix}\text{Mechanical}\\ \text{output}\\ \text{power}\end{pmatrix} + \begin{pmatrix}\text{Rate of increase}\\ \text{in magnetic}\\ \text{energy}\end{pmatrix}. \qquad (3.53)$$

The symbols used in the following theory follow the rules:

(i) Lower case letters are for time-variant variables.

(ii) Upper case letters are for Laplace transforms and stationary values.

(iii) Time-variant torque is denoted by τ, to avoid confusion with t for time.

Now eqn (3.53) is written as

$$-(e_A i_A + e_B i_B) = \tau \frac{d\theta}{dt} + \frac{d}{dt}(\tfrac{1}{2}i_A^2 L_A + \tfrac{1}{2}i_B^2 L_B) \qquad (3.54)$$

where e_A = e.m.f. induced in the A phase

e_B = e.m.f. induced in the B phase

i_A = current in the A phase

i_B = current in the B phase

L_A = inductance of the A phase

L_B = inductance of the B phase

τ = torque developed.

Since we assume that the magnetic circuits are linear and the mutual inductance between the two phases is negligible, the torque can be separated into A and B phase components such that

$$\tau = \tau_A + \tau_B. \qquad (3.55)$$

Hence we have

$$-e_A i_A = \tau_A \frac{d\theta}{dt} + \frac{1}{2}\frac{d}{dt} i_A^2 L_A \qquad (3.56)$$

$$-e_B i_B = \tau_B \frac{d\theta}{dt} + \frac{1}{2}\frac{d}{dt} i_B^2 L_B. \qquad (3.57)$$

3.2.2 *Effect of the permanent magnet on torque production*

The terminal voltage for each phase is the sum of two components; the voltage generated by the permanent-magnet flux linking the phase windings and that caused by current flowing through the phase inductance.

Equation (3.56) for phase A is, therefore, rewritten as

$$-(e_{gA}+e_{LA})i_A = \tau_A \frac{d\theta}{dt} + \frac{1}{2}\frac{d}{dt}i_A^2 L_A \tag{3.58}$$

where e_{LA} is the voltage induced by the current in phase A and is given by

$$e_{LA} = -\frac{d}{dt}(i_A L_A). \tag{3.59}$$

Substituting this into eqn (3.58) we obtain

$$-e_{gA}i_A + i_A \frac{d}{dt}(i_A L_A) = \tau_A \frac{d\theta}{dt} + \frac{1}{2}\frac{d}{dt}i_A^2 L_A. \tag{3.60}$$

Hence we have

$$i_A \frac{d}{dt}(i_A L_A) - \frac{1}{2}\frac{d}{dt}i_A^2 L_A = i_A^2 \frac{dL_A}{dt} + L_A i_A \frac{di_A}{dt} - \frac{1}{2}L_A \frac{di_A^2}{dt} - \frac{1}{2}i_A^2 \frac{dL_A}{dt} \tag{3.61}$$

which, as the second and third terms cancel each other, becomes

$$= \tfrac{1}{2}i_A^2 \frac{dL_A}{dt} = \tfrac{1}{2}i_A^2 \frac{dL_A}{d\theta}\frac{d\theta}{dt}. \tag{3.62}$$

Substituting this relation into eqn (3.58), we get for the phase A torque

$$\tau_A = -e_{gA}i_A/\dot{\theta} + \tfrac{1}{2}i_A^2 \frac{dL_A}{d\theta}, \tag{3.63}$$

where

$$\dot{\theta} = \frac{d\theta}{dt}. \tag{3.64}$$

The second term on the right-hand side of eqn (3.63) represents the torque due to the variation of the phase inductance with rotor position, which is the principle of the VR stepping motors as seen in the theory for the VR motor. In a typical hybrid stepping motor, the variation of the phase inductance is as small as a few percent, and its contribution to the stationary torque is negligible. Hence we obtain for the torque

$$\tau = -(e_{gA}i_A + e_{gB}i_B)/\dot{\theta}. \tag{3.65}$$

The induced voltage for each phase is given by

$$e_{gA} = n(-\phi_1 + \phi_3 - \phi_5 + \phi_7), \tag{3.66}$$

$$e_{gA} = n(-\phi_2 + \phi_4 - \phi_6 + \phi_8) \tag{3.67}$$

where n = number of turns on each pole

ϕ_k = permanent magnet flux in each pole.

It is known from experiments that the waveform of e_g is close to a sine wave, involving some harmonic components.

It we ignore the harmonic components, e_{gA} and e_{gB} are given by

$$e_{gA} = \omega C \cos (\omega t - \rho), \tag{3.68}$$

$$e_{gB} = \omega C \sin (\omega t - \rho) \tag{3.69}$$

where C = constant determined by the motor dimensions and number of turns ($V\,s\,rad^{-1}$)

ρ = a phase anlgle, the 'torque angle' (rad).

The angular frequency ω in these equations is related to the angular speed $\dot{\theta}$ and the number of rotor teeth N_r and is given by

$$\omega = N_r \dot{\theta}. \tag{3.70}$$

3.2.3 *Stationary torque*

Let us assume that the current in each phase is a sinusoidal wave having the same frequency ω as that of the induced voltages:

$$i_A = -I_M \sin \omega t, \tag{3.71}$$

$$i_B = +I_M \cos \omega t. \tag{3.72}$$

Substituting eqns (3.70)–(3.72) into eqn (3.65) we obtain

$$\tau = -\frac{\omega C I_M}{\dot{\theta}} \{\sin (\omega t - \rho) \cos \omega t - \cos (\omega t - \rho) \sin \omega t\}$$

$$\tau = C N_r I_M \sin \rho. \tag{3.73}$$

Since the torque, which is to be balanced with the load, is proportional to $\sin \rho$, the angle ρ is termed the 'torque angle' or 'load angle'.

Thus an ideal hybrid stepping motor generates a rippleless torque, provided that both the induced voltage due to the permanent magnet and the current has a sinusoidal waveform. As in many applications of stepping motors the current waveform is different from the sinusoidal shape, the ripple component in the torque being quite large. Even if a sinusoidally shaped current is applied a torque ripple still appears because of the detenting effect and the harmonics in the induced e.m.f. due to the rotor magnet.

For the study of torque ripple and drives to minimize it see References [5]–[9].

3.3 Tooth structure, number of teeth, steps per revolution, and number of poles

How stepping motors differ from conventional motors was shown in Chapter 2 in terms of application. Seen as a torque production mechanism, the stepping motor's most important characteristic is its teeth. In most other rotating or linear electrical motors, the teeth are not absolutely necessary for torque production but they are inevitably used to minimize the gap between the rotor and stator, while reserving enough space for windings. In stepping motors, however, teeth on both rotor and stator are essential to create torque and to position the rotor at a certain angle. In this section, tooth structure, number of teeth, and their arrangement will be discussed.

3.3.1 *Tooth structure*

The tooth structures of various stepping motors can be divided into three basic types. In the first type, as shown in Fig. 3.14, there are the same number of teeth on the stator and rotor. This is a type which is employed for multi-stack VR motors (see Figs. 2.22 and 2.25). In this structure all teeth are excited and de-energized at the same time. The second type is illustrated in Fig. 3.15 in which the numbers of teeth are different on the stator and rotor. This is the structure employed for single-stack VR motors with a large step angle (see Fig. 2.15, p. 28), and in this machine not all of the teeth are excited at the same time. The third type is illustrated in Fig. 3.16; the stator teeth are arranged in groups on poles, while the rotor teeth are distributed homogeneously on the periphery. This is used in single-stack VR motors with a small step angle and in hybrid stepping motors. In some hybrid stepping motors, there is a slight difference between the rotor tooth-pitch and stator tooth-pitch as shown in Fig. 3.16(b), e.g. 50 tooth-pitch on the rotor and 48 tooth-pitch on the stator.

In designing a stepping motor the determination of the tooth/slot ratio is one of the most important problems, since it strongly affects the static torque characteristics. The tooth/slot ratio also affects the dynamic characteristics, since it is one of the most important factors determining the inductance of each phase. (The relation between the inductance and dynamic characteristics will be discussed in Chapter 4.)

Despite the considerable variety in the designs of machines, all produce torque by doubly-salient VR action, and can be related to a single basic

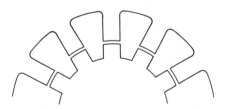

Fig. 3.14. Toothed structure having the same tooth pitch on rotor and stator.

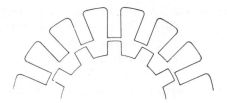

Fig. 3.15. Toothed structure having different tooth pitches on rotor and stator.

magnetic configuration, similar to that of Fig. 3.14. Harris *et al.*[4] carried out an excellent study of static torque production in a tooth structure like this, taking account of saturation. According to their study, the greatest mean torque is produced by the smallest possible gap, and the optimum tooth-width/tooth-pitch ratio ($=t/\lambda$) is theoretically 0.42, independent of size. They say that the t/λ ratio in modern practice ranges from 0.38 to 0.42 and gives some reasons for it.

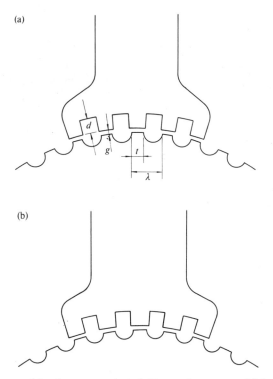

Fig. 3.16. Two types of tooth structure in hybrid stepping motors: (a) has the same tooth pitch on stator and rotor; (b) has a longer tooth pitch on the stator.

Modern hybrid motors and VR motors with the tooth structures of Fig. 3.16 have the following general characteristics.
1. The tooth width/tooth pitch ratio is near to 0.5.
2. The slot depth (d) in the stator is about a half of the tooth pitch λ.
3. The t/λ ratio for the rotor ranges 0.25 to 0.45.
4. The slot shape is semicircular for the rotor and either rectangular or semicircular for the stator.
5. The gap length g is taken to be as small as possible, taking account of mass production techniques, and ordinarily 0.1 mm, but in some special cases 0.02 mm.

3.3.2 *Relation between number of teeth, steps per revolution, and number of phases*

The number of phases m, number of rotor teeth N_r, and steps per revolution S, are related by the general equation[10]

$$S = mN_r. \tag{3.74}$$

This is true with one-phase-on or two-phase-on operation, but for the half-step drive we must use

$$S = 2mN_r. \tag{3.75}$$

Let us suppose that an excitation sequence is started from a phase in an m-phase motor. When a cycle of the excitation sequence has been completed and the first phase is again excited, having made m steps, the motor will have rotated one tooth-pitch. This situation is illustrated in Fig. 2.13 (p. 28) on a single-stack three-phase motor having four teeth on the rotor. Since m pulses are transmitted to rotate the motor by one tooth pitch, mN_r pulses are needed to complete one revolution. This explanation also applies to a multi-stack motor.

Care is needed when considering the single-stack bifilar-wound motor. As explained in Section 2.3.4, the three-phase VR motor with bifilar windings may be regarded as a six-phase motor. But as is obvious from Fig. 2.60, this type of motor proceeds one tooth pitch in three steps. Therefore this motor should be treated as a three-phase motor in calculating the steps per revolution.

With regard to the tooth structure of the second type, in which the stator and rotor have different tooth pitches as shown in Fig. 3.15, we have the relation

$$q = |N_r - N_s| \tag{3.76}$$

where N_r = number of rotor teeth
N_s = number of stator teeth

$$q = \frac{N_s}{m} \text{ number of stator teeth per phase.} \tag{3.77}$$

Eliminating m from eqns (3.74) and (3.77) we obtain

$$S = \frac{N_r N_s}{q}. \tag{3.78}$$

Furthermore, considering eqn (3.76), we get

$$S = \frac{N_r N_s}{N_r - N_s}. \tag{3.79}$$

From eqns (3.76), (3.77), and (3.79), we obtain for the relation between q, S, and m

$$S = m(m+1)q \quad \text{for } N_r > N_s \tag{3.80}$$

$$S = m(m-1)q \quad \text{for } N_r < N_s. \tag{3.81}$$

We can verify that the largest step angle is 30° as shown in Fig. 2.12 for this type of motor as follows. The minimum q is 2, since at least two magnetic poles are needed in the heteropolar structure. The minimum m is 3 in order that the rotational direction is determined by the sequence of excitation. Then we have from eqn (3.81)

$$S = 3 \times 2 \times 2 = 12 \tag{3.82}$$

and the step angle θ_s is

$$\theta_s = \frac{360°}{S} = \frac{360°}{12} = 30°. \tag{3.83}$$

References for Chapter 3

[1] Seely, S. (1962). *Electromechanical energy conversion.* p. 19. McGraw Hill Book Company, Inc.
[2] Byrne, J. V. (1972). Tangential forces in overlapped pole geometries, incorporating ideally saturable material. *Trans IEEE Mag.* **8,** (1), 2–9.
[3] Lawrenson, P. J., Hodson, D. P., and Harris, M. R. (1976). Electromagnetic forces in saturated magnetic circuits. *Proc. Conference on small electrical machines.* Institution of Electrical Engineers, London, pp. 89–92.
[4] Harris, M. R., Hughes, A., and Lawrenson, P. J. (1975). Static torque production in saturated doubly-salient machines. *Proc. IEE* **122,** (10), 1121–7.
[5] Patterson, M. L. (1977). Analysis and correction of torque harmonics in permanent-magnet step motors. *Proc. Sixth annual symposium on Incremental motion control systems and devices.* Department of Electrical Engineering, University of Illinois, pp. 25–37.
[6] Patterson, M. L. and Haselby, R. D. (1977). A microstepped XY controller with adjustable phase current waveforms. Ibid., pp. 163–8.
[7] Pritchard, E. K. Mini-stepping observations on stepping motors. Ibid., pp. 169–78.

[8] Layer, H. P. (1977). Digital sine–cosine mini-stepping motor drive. Ibid., pp. 179–81.

[9] Kenjo, T. and Takahashi, H. (1979). Speed ripple characteristics of hybrid stepping motors driven in the ministep mode. *Proc. International conference of stepping motors and systems.* University of Leeds, pp. 87–93.

[10] Kuo, B. C. (1979). *Incremental motion control. Step motors and control systems.* p. 15. SRL Publishing Company, Champaign, Illinois.

4. Fundamental theory of the dynamic characteristics of stepping motors

The dynamic characteristics are, as well as the static characteristics, important and must be taken into account in putting a stepping motor in a system. A number of theories of the dynamic performance of stepping motors have been reported by many authors. Among them, the theory developed by Lawrenson and Hughes, University of Leeds, is selected to present here, since their theory is systematic and covers most essential aspects of the dynamic behaviour of modern stepping motors.

The main symbols used in this chapter and Chapter 3 are shown in Table 3.1 on p. 67.

4.1 Fundamental equations

The model[1] of Fig. 4.1 is used here to analyse an oscillatory phenomenon and electromagnetic damping in stepping motors. Figure 4.1(a) is a model of a PM motor, but it is applicable to a hybrid stepping motor; (b) is for the single-stack VR motor. In those models the two phases are indicated by A and B. The rotor in model (a) has $2p$ magnetic poles, and that in model (b) has $2p$ teeth, while the stator has a set of identical poles and windings equally arranged at intervals of λ.

4.1.1 Permanent-magnet and hybrid motors

If the peak flux linkage produced by the permanent magnet is $n\Phi_M$, the torque produced by a current i_A in winding A is given by

$$\tau_A = -pn\Phi_M i_A \sin p\theta. \tag{4.1}$$

Similarly, the torque developed by the current i_B is given by

$$\tau_B = -pn\Phi_M i_B \sin p(\theta - \lambda) \tag{4.2}$$

where p is the number of pairs of magnetic poles, but it may be replaced by N_r (the number of rotor teeth) in the case of the hybrid motor illustrated by the model of Fig. 4.2.

Since in those models the centre of the pole A is the origin of θ, the flux linkage $n\phi$ is approximately shown by the sinusoidal space distribution

$$n\phi = n\Phi_M \cos p\theta. \tag{4.3}$$

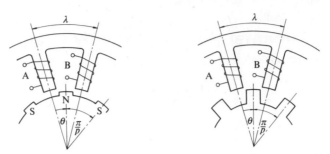

Fig. 4.1. Models for stepping motors used for analysis of electromagnetic damping (after Ref. [1]).

The e.m.f. induced in phase A is

$$e_{gA} = -n\frac{d\phi}{dt} = -n\frac{d\phi}{d\theta}\frac{d\theta}{dt} = (np\Phi_M \sin p\theta)\dot{\theta}. \tag{4.4}$$

Comparing eqns (4.1) and (4.4), we find that

$$\tau_A = -e_{gA}i_A/\dot{\theta},$$

which is the same form as the first term in eqn (3.63). The second term representing the reluctance torque is ignored here.

The equation of motion of the rotor is

$$J\frac{d^2\theta}{dt^2} + D\frac{d\theta}{dt} + pn\Phi_M i_A \sin p\theta + pn\Phi_M i_B \sin p(\theta - \lambda) = 0. \tag{4.5}$$

Here D denotes the viscous damping coefficient which accounts for the presence of air and friction, and for the second order electromagnetic effects arising from hysteresis and eddy currents. The voltage equations

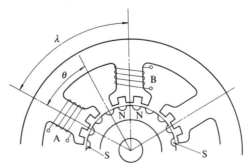

Fig. 4.2. Model for a hybrid stepping motor; $p = N_r$.

for the stator windings are

$$V - ri_A - L\frac{di_A}{dt} - M\frac{di_B}{dt} + \frac{d}{dt}(n\Phi_M \cos p\theta) = 0, \qquad (4.6)$$

$$V - ri_B - L\frac{di_B}{dt} - M\frac{di_A}{dt} + \frac{d}{dt}\{n\Phi_M \cos p(\theta - \lambda)\} = 0 \qquad (4.7)$$

where V = DC terminal voltage
$\quad\quad L$ = Self-inductance of each phase
$\quad\quad M$ = Mutual inductance
$\quad\quad r$ = Stator-circuit resistance.

In these equations it is assumed that L and M are independent of θ. What we are going to discuss is the difference between the single-phase excitation and the two-phase excitation in the dynamic behaviour. Since the above equations apply when both A and B phases are excited, they can deal with the problem of the two-phase excitation. One of the features of Lawrenson and Hughes' method is that the single-phase excitation can be dealt with, too, only by putting $\lambda = 0$ in the final results or at an earlier stage in the process of analysis. That is, the single-phase excitation corresponds to the case that both poles come at the same place (are coincident).

The equations (4.5)–(4.7) are non-linear differential equations. Since it is very difficult to deal with a non-linear differential equation fully analytically, we will start by linearizing the equations. If the windings of the two phases carry the stationary current I_0 in a direction to create the south pole, the equilibrium position is at $\theta = \lambda/2$ in Fig. 4.3. The disturbance from the equilibrium position is denoted by $\delta\theta$, which is a function of time t but small in magnitude in the following analyses. When the rotor revolves or oscillates, the currents in both windings will deviate from the

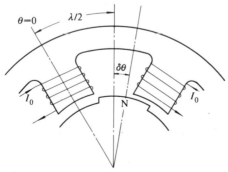

Fig. 4.3. Position control of rotor.

stationary value by $\delta i_A(t)$ and $\delta i_B(t)$, respectively. Equations (4.5) and (4.6) are linearized by the following procedure.

We put

$$\theta = \frac{\lambda}{2} + \delta\theta \tag{4.8}$$

$$i_A = I_0 + \delta i_A \tag{4.9}$$

$$i_B = I_0 + \delta i_B. \tag{4.10}$$

Then we have

$$\sin p\theta = \sin \left(\frac{p\lambda}{2} + p\delta\theta\right) = \sin \frac{p\lambda}{2} \cos p\delta\theta + \cos \frac{p\lambda}{2} \sin p\delta\theta. \tag{4.11}$$

Since when $p\delta\theta$ is a small angle

$$\cos p\delta\theta \simeq 1 \tag{4.12}$$

$$\sin p\delta\theta \simeq p\delta\theta. \tag{4.13}$$

Equation (4.11) is simplified as

$$\sin p\theta = \sin \frac{p\lambda}{2} + p \cos \frac{p\lambda}{2} (\delta\theta). \tag{4.14}$$

Similarly

$$\sin p(\theta - \lambda) = \sin p\left(\frac{\lambda}{2} + \delta\theta - \lambda\right) = -\sin p(\lambda - \delta\theta)$$

$$= -\sin \frac{p\lambda}{2} + p\left(\cos \frac{p\lambda}{2}\right)(\delta\theta) \tag{4.15}$$

By substituting eqns (4.8), (4.9), (4.10), (4.14), and (4.15) into (4.5), we obtain

$$J\frac{d^2(\delta\theta)}{dt^2} + D\frac{d(\delta\theta)}{dt} + pn\Phi_M(I_0 + \delta i_A)\left\{\sin \frac{p\lambda}{2} + p\left(\cos \frac{p\lambda}{2}\right)(\delta\theta)\right\} -$$

$$- pn\Phi_M(I_0 + \delta i_B)\left\{\sin \frac{p\lambda}{2} - p\left(\cos \frac{p\lambda}{2}\right)(\delta\theta)\right\} = 0. \tag{4.16}$$

If we ignore the products of disturbances, e.g. $\delta i_A \delta\theta$, we obtain the linear differential equation

$$J\frac{d^2(\delta\theta)}{dt^2} + D\frac{d(\delta\theta)}{dt} + 2p^2\Phi_M nI_0\left(\cos \frac{p\lambda}{2}\right)(\delta\theta) +$$

$$+ p\Phi_M n\left(\sin \frac{p\lambda}{2}\right)(\delta i_A - \delta i_B) = 0. \tag{4.17}$$

Following a similar procedure, eqns (4.6) and (4.7) are linearized as

$$r(\delta i_A) + L\frac{d(\delta i_A)}{dt} + M\frac{d(\delta i_B)}{dt} - p\Phi_M n \sin\left(\frac{p\lambda}{2}\right)\frac{d(\delta\theta)}{dt} = 0 \qquad (4.18)$$

$$r(\delta i_B) + L\frac{d(\delta i_B)}{dt} + M\frac{d(\delta i_A)}{dt} + p\Phi_M n \sin\left(\frac{p\lambda}{2}\right)\frac{d(\delta\theta)}{dt} = 0. \qquad (4.19)$$

Let us consider which kind of function expresses the rotor position $\delta\theta(t)$ after the rotor is released from rest at an angle θ_i from the equilibrium position. Moreover, how currents i_A and i_B behave is another interesting problem. To determine these, we must solve these equations with initial conditions $\delta\theta = \theta_i$ and $d(\delta\theta)/dt = 0$ at $t = 0$. We solve the equations by means of the Laplace transformation, putting $d/dt = s$ and $d^2/dt^2 = s^2$. The Laplace transforms are indicated by capital letters as follows:

$$\delta\theta(t) \rightarrow \Theta(s)$$
$$\delta i_A(t) \rightarrow I_A(s)$$
$$\delta i_B(t) \rightarrow I_B(s).$$

The solutions are

$$I_A = -I_B = \frac{p\Phi_M n \sin\left(\frac{p\lambda}{2}\right)(s\Theta - \theta_i)}{(r + L_p s)}, \qquad (4.20)$$

$$\Theta = \frac{\left\{s^2 + \left(\frac{r}{L_p} + \frac{D}{J}\right)s + \left(\frac{r}{L_p}\frac{D}{J} + k_p\omega_{np}^2\right)\right\}\theta_i}{s^3 + \left(\frac{r}{L_p} + \frac{D}{J}\right)s^2 + \left\{\frac{r}{L_p}\frac{D}{J} + \omega_{np}^2(1 + k_p)\right\}s + \left(\frac{r}{L_p}\right)\omega_{np}^2} \qquad (4.21)$$

where

$$L_p = L - M, \qquad (4.22)$$

$$k_p = \frac{n\Phi_M}{L_p I_0}\frac{\sin^2\left(\frac{p\lambda}{2}\right)}{\cos\left(\frac{p\lambda}{2}\right)}, \qquad (4.23)$$

$$\omega_{np}^2 = \frac{2p^2\Phi_M n I_0 \cos\left(\frac{p\lambda}{2}\right)}{J}. \qquad (4.24)$$

Equation (4.21) which governs the behaviour of $\theta(t)$ is an important equation here. The most remarkable feature of this equation is that its denominator is of the third order with respect to s, the physical meanings of which will be discussed later. Equation (4.20) indicates that the transient currents in phases A and B are equal but opposite to each other. Because of this character, the two-phase excitation scheme offers excellent damping, as stated later on. Before proceeding to the solution in the time domain, it is shown that essentially identical equations will occur for the VR motor.

4.1.2   VR motor

The following treatment is based on the single-stack VR motor, but the results can be applied to multistack types simply by setting all the mutual inductances to zero. The self and mutual inductances of the two-phase windings in the model of Fig. 4.1(b) have the general forms.

$$L_A = L_0 + L \cos 2p\theta \tag{4.25}$$

$$L_B = L_0 + L \cos 2p(\theta - \lambda) \tag{4.26}$$

$$M_{AB} = -M_0 + M \cos 2p\left(\theta - \frac{\lambda}{2}\right). \tag{4.27}$$

The minus sign for M_0 in eqn (4.27) indicates that positive current in one winding produces negative flux linkages in the other. The torque produced by currents i_A and i_B, after eqn (3.52) is given by

$$\tau = \tfrac{1}{2}i_A^2 \frac{dL_A}{d\theta} + \tfrac{1}{2}i_B^2 \frac{dL_B}{d\theta} + i_A i_B \frac{dM_{AB}}{d\theta} \tag{4.28}$$

$$= -\left\{ i_A^2 pL \sin 2p\theta + i_B^2 pL \sin 2p(\theta - \lambda) + 2i_A i_B pM \sin 2p\left(\theta - \frac{\lambda}{2}\right) \right\} \tag{4.29}$$

and the equation of motion is thus

$$J\frac{d^2\theta}{dt^2} + D\frac{d\theta}{dt} + i_A^2 pL \sin 2p\theta + i_B^2 pL \sin 2p(\theta - \lambda) +$$

$$+ 2i_A i_B pM \sin 2p(\theta - \lambda/2) = 0. \tag{4.30}$$

The voltage equations for the two windings are

$$V - ri_A - \frac{d}{dt}(L_A i_A) - \frac{d}{dt}(M_{AB} i_B) = 0 \tag{4.31}$$

$$V - ri_B - \frac{d}{dt}(L_B i_B) - \frac{d}{dt}(M_{AB} i_A) = 0. \tag{4.32}$$

These equations are linearized as

$$J\frac{d^2(\delta\theta)}{dt^2} + D\frac{d(\delta\theta)}{dt} + 4p^2 I_0^2 (M + L \cos p\lambda)(\delta\theta) +$$

$$+ 2pI_0 L \sin p\lambda (\delta i_A - \delta_B) = 0, \tag{4.33}$$

$$r(\delta i_A) + (L_0 + L \cos p\lambda)\frac{d}{dt}(\delta i_A) + (M - M_0)\frac{d}{dt}(\delta i_B) -$$

$$- 2pI_0 L \sin p\lambda \frac{d}{dt}(\delta\theta) = 0, \tag{4.34}$$

$$r(\delta i_B) + (L_0 + L \cos p\lambda)\frac{d}{dt}(\delta i_B) + (M - M_0)\frac{d}{dt}(\delta i_A) +$$

$$+ 2pI_0 L \sin p\lambda \frac{d}{dt}(\delta\theta) = 0. \tag{4.35}$$

Equations (4.33) to (4.35) are identical in form, in the variables $\delta\theta$, δi_A, and δi_B, to eqns (4.17) to (4.19) for the PM motor, and the solutions of the equations, subject to the same initial conditions, are

$$I_A = -I_B = \frac{2pI_0 L \sin (p\lambda)(s\Theta - \theta_i)}{(r + L_v s)}, \tag{4.36}$$

$$\Theta(s) = \frac{\left\{s^2 + \left(\frac{r}{L_v} + \frac{D}{J}\right)s + \frac{r}{L_v}\frac{D}{J} + k_p\omega_{nv}^2\right\}\theta_i}{s^3 + \left(\frac{r}{L_v} + \frac{D}{J}\right)s^2 + \left\{\frac{r}{L_v}\frac{D}{J} + \omega_{nv}^2(1 + k_v)\right\}s + \left(\frac{r}{L_v}\right)\omega_{nv}^2} \tag{4.37}$$

where

$$L_v = L_0 + L \cos p\lambda - M + M_0, \tag{4.38}$$

$$k_v = \frac{2L^2 \sin^2 p\lambda}{L_v(M + L \cos p\lambda)}, \tag{4.39}$$

$$\omega_{nv}^2 = \frac{4p^2 I_0^2(M + L \cos p\lambda)}{J}. \tag{4.40}$$

4.2 Transfer functions of stepping motors

Equations (4.21) and (4.37) are the expressions in the s domain for the transient locus of $\delta\theta$ with the initial position $\delta\theta = \theta_i$. Before discussing the damping characteristics, let us define the transfer function of stepping motors.

Stepping motors are usually employed for position control. It is expected in Fig. 4.3 that the centre of the rotor magnetic pole is positioned at $\theta = \lambda/2$ by exciting the two phases equally; the demanded value in the present case is $\theta = \lambda/2$. Let the Laplace transform of the demanded value be Θ_i, and let the Laplace transform of the actual position $\delta\theta(t)$ be Θ_0. Then the transfer function is defined as

$$G(s) = \frac{\Theta_0}{\Theta_i} \tag{4.41}$$

and we shall discuss this equation for various cases.

4.2.1 Single-phase operation

As stated before, when $\lambda = 0$, the windings behave as one. Hence the solutions for single-phase operation can be obtained simply by putting $\lambda = 0$ in the two-phase equations. The final terms in each of eqns (4.17) to (4.19) and (4.30) to (4.32) then become zero, and the mechanical and electrical equations become independent. The transfer equation for the PM motor is derived only from eqn (4.17) and similarly for the VR motor only from eqn (4.32). In order to proceed in the PM motor case, putting $\lambda = 0$ in eqn (4.17), we get

$$J\frac{d^2(\delta\theta)}{dt^2} + D\frac{d(\delta\theta)}{dt} + 2p^2\Phi_M nI_0\delta\theta = 0. \qquad (4.42)$$

Since $\delta\theta$ is the actual position θ_0 less the demanded position θ_i:

$$\delta\theta = \theta_0 - \theta_i \qquad (4.43)$$

eqn (4.42) is rewritten to be

$$J\frac{d^2\theta_0(t)}{dt^2} + D\frac{d\theta_0(t)}{dt} + 2p^2\Phi_M nI_0\theta_0(t) = 2p^2\Phi_M nI_0\theta_i. \qquad (4.44)$$

The Laplace transform of this equation with initial conditions $\theta_0 = 0$, $d\theta_0/dt = 0$ at $t = 0$ is

$$(s^2J + sD + 2p^2\Phi_M nI_0)\Theta_0(s) = 2p^2\Phi_M nI_0\Theta_i \qquad (4.45)$$

from which the following transfer function is obtained

$$G(s) = \frac{\Theta_0}{\Theta_i} = \frac{2p^2\Phi_M nI_0}{Js^2 + Ds + 2p^2\Phi_M nI_0} = \frac{\omega_{np}^2}{s^2 + Ds/J + \omega_{np}^2}. \qquad (4.46)$$

Hence ω_{np} is the 'natural angular frequency' which is given by

$$\omega_{np} = \sqrt{(2p^2\Phi_M nI_0/J)}, \qquad (4.47)$$

or

$$\omega_{np} = \sqrt{(N_r K_T I_0/J)}. \qquad (4.48)$$

The form of eqn (4.48) is the well-known form for the hybrid stepping motor. K_T in eqn (4.48) is the 'torque constant' and given by

$$K_T = 2nN_r\Phi_M. \qquad (4.49)$$

Equation (4.49) is derived as follows. The total torque τ is the sum of τ_A of eqn (4.1) and τ_B of eqn (4.2), but $\lambda = 0$ for the single-phase excitation and $p = N_r$ for the hybrid motor. Consequently

$$\tau = -2N_r n\Phi_M I_0 \sin(N_r\delta\theta). \qquad (4.50)$$

If the rotor position $\delta\theta$ is near zero, $\sin(N_r\delta\theta) \simeq N_r\delta\theta$, and we obtain

$$\tau = -2N_r^2 n\Phi_M I_0 \, \delta\theta. \tag{4.51}$$

When the torque produced in a stepping motor is expressed as a linear function of $\delta\theta$, it is often expressed in the form

$$\tau = -K_T N_r I_0 \, \delta\theta. \tag{4.52}$$

From comparison of eqns (4.50) and (4.52) it is found that the torque constant K_T in eqn (4.52) must have the form of eqn (4.49). Replacing p with N_r in eqn (4.47) and eliminating n by the use of eqn (4.49) we attain the form of eqn (4.48) for ω_{np}. Since the number of turns is $2n$ here, if it is replaced by n_1 being the number of turns in a phase in the normal motor, K_T is expressed in the form

$$K_T = n_1 N_r \Phi_M. \tag{4.53}$$

The torque constant is theoretically expressed in this form, but it is ordinarily obtained by approximation of the T/θ characteristic curve by a linear line as explained in Fig. 4.4.

A transfer function of the form of eqn (4.46) can also be derived from eqn (4.21). Since $\lambda = 0$ in this case the mutual inductance M is the same as the self-inductance L, and so L_p of eqn (4.22) becomes zero. After multiplying both numerator and denominator by L_p/r, we put $L_p = 0$ to obtain

$$\Theta = \frac{(s + D/J)\theta_i}{s^2 + (D/J)s + \omega_{np}^2}. \tag{4.54}$$

If we replace the numerator by a constant which makes the value of the function unity when $s = 0$, it is the transfer function. This is obviously the same as eqn (4.46). The transfer function for the VR motor can be determined by a similar procedure.

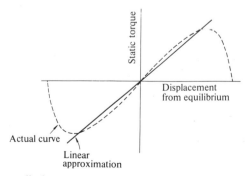

Fig. 4.4. Torque versus displacement curve and its linear approximation.

4.2.2   Current-source drive

There are several choices for the waveforms of the excitation voltage or current. In eqns (4.6) and (4.7) or (4.31) and (4.32) a constant voltage is applied, which implies a voltage-source drive. When the torque developed in a hybrid stepping motor was discussed in Section 3.2, however, the currents were of sinusoidal form. To specify current means that the motor is driven from a current source. The transfer function for the current-source drive will be discussed here. The torque given by eqn (3.73) is reproduced here

$$\tau = C N_r I_M \sin \rho \qquad (4.55)$$

where C = constant determined by the dimensions of the motor and the number of turns, being of the same dimension as the torque constant (cf. eqn (4.52))

I_M = peak current.

The torque angle ρ has the meaning

$$\rho = \xi_i - \xi_0 \qquad (4.56)$$

where ξ_i = electrical angle of the demanded rotor position
ξ_0 = electrical angle of the actual rotor position.

It is reasonable to treat ωt in eqns (3.71) and (3.72) as the demanded position in the position control using the two-phase sinusoidal current drive. Therefore

$$\xi_i = \omega t. \qquad (4.57)$$

This means that the demanded position is varying at a rate ω, or it may alternatively be said that the velocity is commanded. If ξ_i is fixed at ωt_1 at $t = t_1$, the rotor is expected to fall at the position which makes $\rho = 0$, or $\xi_0 = \xi_i$. (The friction load is neglected.) The ω in eqn (4.57) is equal to the angular frequency of the induced voltage given by eqns (3.68) and (3.69) and related to the number of teeth N_r in such a way that ω is proportional to N_r if the speed is fixed. Hence the electric angles ξ_i and ξ_0 are related to the mechanical angles θ_i and θ_0, respectively as follows:

$$\theta_i = \xi_i/N_r \qquad (4.58)$$

$$\theta_0 = \xi_0/N_r. \qquad (4.59)$$

Then the torque of eqn (4.55) is written as

$$\tau = C N_r I_M \sin N_r(\theta_i - \theta_0)$$
$$= K_T I_M N_r \sin N_r(\theta_i - \theta_0) \qquad (4.60)$$

where the torque constant K_T is

$$K_T = C N_r. \qquad (4.61)$$

If $\theta_i \simeq \theta_0$

$$\tau \simeq K_T I_M N_r (\theta_i - \theta_0). \tag{4.62}$$

The equation of motion is

$$J\frac{d^2\theta_0}{dt^2} + D\frac{d\theta_0}{dt} = K_T I_M N_r (\theta_i - \theta_0). \tag{4.63}$$

The Laplace transform with initial conditions being zero is

$$(Js^2 + Ds + K_T I_M N_r)\Theta(s) = K_T I_M N_r \Theta_i. \tag{4.64}$$

From this equation the transfer function is derived as

$$G(s) = \frac{\Theta_0}{\Theta_i} = \frac{\omega_{np}^2}{s^2 + (D/J)s + \omega_{np}^2}. \tag{4.65}$$

It should be noted that the transfer function of the stepping motor driven in the current-source two-phase excitation is the same as that for the voltage-source single-phase excitation. The transfer function can be derived also from eqn (4.21). The current-source drive employs a power supply with infinitive internal impedance, or with a large resistance connected in series with the phase winding. After multiplying both numerator and denominator in eqn (4.21) by r/L, we set r to ∞ to obtain

$$\Theta = \frac{\left(s + \dfrac{D}{J}\right)\theta_i}{s^2 + \dfrac{D}{J}s + \omega_{np}^2}. \tag{4.66}$$

Replacing the numerator by a constant which makes the value of the equation unity for $s = 0$, we obtain the transfer function, eqn (4.65). A similar function will be obtained for the VR motor.

4.2.3   Two-phase excitation in the voltage-source drive

The transfer function is derived from eqn (4.21) as

$$G(s) = \frac{(r/L_p)\omega_{np}^2}{s^3 + \left(\dfrac{r}{L_p} + \dfrac{D}{J}\right)s^2 + \left\{\dfrac{r}{L_p}\dfrac{D}{J} + \omega_{np}^2(1 + k_p)\right\}s + \left(\dfrac{r}{L_p}\right)\omega_{np}^2}. \tag{4.67}$$

A similar form is obtained for the VR type.

4.3   Single-step response

We are ready to consider the transient response of a stepping motor. We shall compare the case of the second-order transfer function and the case of the third-order transfer function. This becomes the comparison of the single-phase drive and the two-phase excitation drive.

4.3.1 Second-order transfer function

The transfer function for the single-phase drive and the current-source drive is written in the form

$$G(s) = \frac{\Theta_0}{\Theta_i} = \frac{\omega_n^2}{s^2 + 2\zeta\omega_n s + \omega_n^2} \qquad (4.68)$$

where ζ is the 'damping ratio' and has the form

$$\zeta = D/2J\omega_n. \qquad (4.69)$$

The form of eqn (4.68) is a well-known form in the feedback-control theory. The indicial response, or the response of θ_0 to the step function of θ_i, has different trajectories depending on the size of ζ as shown in Fig. 4.5. Generally, the response is oscillatory for $\zeta < 1$. Since the damping ratio of an ordinary stepping motor is less than 0.5, the indicial response is oscillatory. The time from the initiation of the step-up to the instant beyond which the rotor is always within a limit is referred to as the settling time. The settling time is different depending on the limit; e.g. as shown in Fig. 4.6 the settling time of the ±2 per cent limit is longer than the ±5 per cent one.

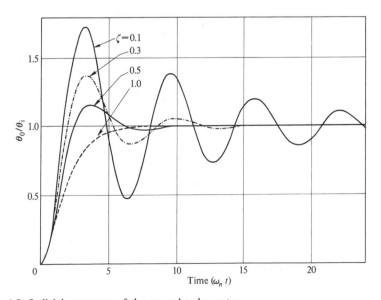

Fig. 4.5. Indicial responses of the second-order system.

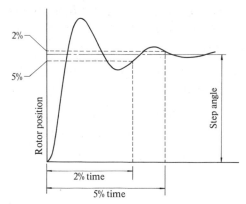

Fig. 4.6. Examples of settling time: the ordinary 5 per cent settling time and the 2 per cent one as a special case.

4.3.2 Third-order transfer function

In the two-phase-on drive the equation of motion and the voltage equation are dependent on each other. The solution eqn (4.20) or eqn (4.36) shows that the motional e.m.f.s produce equal and opposite currents in the excited phases. These currents together constitute, in effect, a circulating current superimposed on the original current I_0. The total current drawn from the supply remains steady at $2I_0$, and there is therefore no change in the power supplied by the source. If the rotor motion is oscillatory, the circulating current will also oscillate and generate Joule heating in the windings. Since this phenomenon indicates that the kinetic energy of the rotor is absorbed to be converted into heat loss, the rotor oscillation may be damped out quickly. This is the explanation of the electromagnetic damping in the two-phase-on operation. The following is the analytic treatment made by Lawrenson and Hughes.

The denominator of a transfer function put equal to zero is called the 'characteristic equation'. When the damping constant D due to air resistance, etc. is assumed to be zero, the characteristic equation of the two-phase-on drive is

$$s^3 + \left(\frac{r}{L}\right)s^2 + (1+k)\omega_n^2 s + \left(\frac{r}{L}\right)\omega_n^2 = 0. \tag{4.70}$$

The parameters in this equation have the following physical meanings:

$r/L =$ the inverse of the effective time constant of the circuit

$k =$ dimensionless constant to give a measure of the inherent damping potential of the motor

$\omega_n =$ undamped natural frequency of small oscillations about equilibrium.

This cubic equation has one real root and a pair of imaginary conjugate roots for $0 < k < 8$; and, since most motors have values of k less than 1, the equation can be expressed in the form

$$(s + \alpha)\{(s + \beta)^2 + \omega^2\} = 0. \tag{4.71}$$

It follows from eqn (4.71) that the time response will always be of the form

$$\theta(t) = Ae^{-\alpha t} + Be^{-\beta t} \cos(\omega t - \gamma) \tag{4.72}$$

where A, B, and γ are constants determined from the initial condition.

It can be said from eqn (4.72) that rapid settling to the equilibrium point under all conditions is possible if α and β are both large. Parameters α, β, and ω are related to the system parameters r, L, k, ω_n by the equations

$$\alpha + 2\beta = r\,L,$$
$$\alpha(\beta^2 + \omega^2) = (r\,L)\omega_n^2, \tag{4.73}$$
$$\beta^2 + \omega^2 + 2\alpha\beta = (1 + k)\omega_n^2.$$

Those equations are graphically expressed in Fig. 4.7. It is seen that β/ω_n is maximum at a $(r/L)\omega_n$ for each k. The relations as to the maximum values are as follows:

$$\text{maximum value of } \beta/\omega_n = k/4 \tag{4.74}$$

$$\text{value of } (r/L)\omega_n = 1 + \frac{k}{2} \tag{4.75}$$

$$\text{corresponding value of } \alpha/\omega_n = 1. \tag{4.76}$$

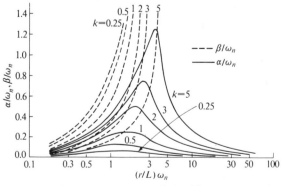

Fig. 4.7. Variation of damping parameters α and β with motor parameter k and circuit parameter r/L (after Ref. [1], reproduced by permission of the Institution of Electrical Engineers, and by courtesy of Professor P. J. Lawrenson and Dr A. Hughes.)

For most modern motors k is less than unity and β is therefore less than $\alpha/4$; the first term decays at least four times as fast as the second in eqn (4.69). Thus it is β that dominates the settling time.

4.3.3 *Effects of the series resistance(s)*

It was shown that the circuit resistance to minimize the settling time for the two-phase excitation is given by eqn (4.75). This is true with both the VR and PM motor. Figure 4.8 shows indicial responses measured on a multistack three-phase VR motor. When an external 32 Ω resistor is connected to each phase whose resistance is 0.5 Ω, the settling time is minimum. The single-phase drive results are such as shown in Fig. 4.9, and are in complete contrast with the two-phase-on response. Firstly the damping is very much lower, and secondly, the value of the stator circuit resistance has no discernible effect on the damping. This principle of attaining good damping by optimum design of winding impedance or by addition of a proper resistance to the winding was patented by Lawrenson and Hughes.[2]

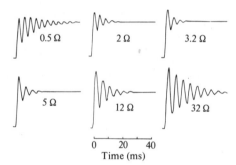

Fig. 4.8. Measured position–time responses on a three-phase VR motor in the two-phase-on excitation mode (after Ref. [1], reproduced by permission of the Institution of Electrical Engineers, and by courtesy of Professor P. J. Lawrenson and Dr A Hughes.)

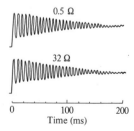

Fig. 4.9. Measured position–time responses in the single-phase-on mode (after Ref. [1], reproduced by permission of the Institution of Electrical Engineers, and by courtesy of Professor P. J. Lawrenson and Dr A Hughes).

To have the maximum starting frequency and the maximum slewing rate improved in the two-phase-on drive, it is necessary to use high values for the external resistors to minimize the electrical time constant L/r. Hence a compromise should be reached between good damping and high stepping rate. To have both good torque–speed characteristics and damping some measures must be taken in the driving circuit.

4.3.4 *Comparison of PM and VR motors*

It is widely believed that VR stepping motors are inferior to PM motors with respect to damping, because they have no magnet. But there seems to be no definite explanation why the parameter k_p of a PM motor is necessarily much larger than k_v of a VR motor. Figure 4.10 shows single-step responses measured in a hybrid motor (which is a special type of PM motor) and a VR motor with step angle of 1.8°.[3] Both motors are of the same size, and the measurements were done with the same m.m.f. using the circuit of Fig. 5.21 with $C = 1\ \mu F$, $E = 24\ V$, and R_e to give a phase current of 1.5 A. The settling time is a little longer with the VR motor than with the hybrid motor.

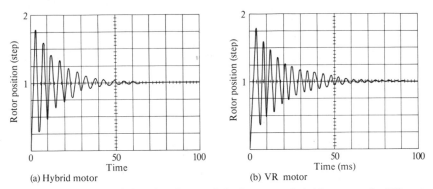

(a) Hybrid motor   (b) VR motor

Fig. 4.10. Comparison of damping characteristics between a hybrid motor and a VR motor in the two-phase-on mode (after Ref. [3]). Test machines: Sanyo Step-Syn 103–775–6 for hybrid type and its modification for VR type. Drive circuit: Fig. 5.21 (p. 135); $E = 24\ V$, $C = 1.0\ \mu F$, R_e is adjusted to have 1.5 A per phase.

4.4 Torque versus speed characteristics

To characterize the torque versus speed relations of a stepping motor the graph as shown in Fig. 4.11 is presented. The ordinate is the torque and the abscissa is the stepping rate. The stepping rate is proportional to speed, but the proportional constant or the steps per revolution depends on tooth structure, number of phases, and excitation mode. The two curves in the figure are the pull-in torque curve and pull-out torque curve which is known also as the slewing curve.

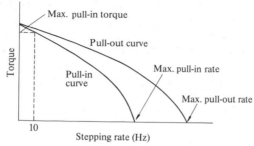

Fig. 4.11. Torque versus speed characteristic curves.

4.4.1 *Pull-out torque characteristic(s)*

The pull-out torque versus speed curve represents the maximum friction-torque load that a stepping motor can drive before losing synchronism at a specified stepping rate. A simple but neat theory was proposed by Hughes *et al.*[4] for the pull-out characteristics for the two-phase hybrid motor. The outline of this theory is presented here. From eqns (4.1) and (4.2), putting $p\lambda = \pi/2$ because we are considering a two-phase motor, the expression for the torque is derived in the form

$$\tau = -nN_r\Phi_M\{i_A \sin(N_r\theta) + i_B \cos(N_r\theta)\}. \tag{4.77}$$

Neglecting the mutual inductance M, the voltages of eqns (4.6) and (4.7) are expressed as

$$v_A = ri_A + L\frac{di_A}{dt} + \frac{d}{dt}(n\Phi_M \cos N_r\theta) \tag{4.78}$$

$$v_B = ri_B + L\frac{di_B}{dt} - \frac{d}{dt}(n\Phi_M \sin N_r\theta). \tag{4.79}$$

The resistance r is the combined resistance of the windings together with external forcing resistance; and the self-inductance L is taken to be independent of rotor position, which is a fairly reasonable assumption for hybrid motors.

The voltage waveform in many applications, e.g. in the bridge-circuit drive shown in Fig. 5.28 (p. 139), is square wave as shown in Fig. 4.12. For the sake of simplicity in analysis, however, the phases are assumed to be supplied from a two-phase sinusoidal supply, so that the phase voltages are given by

$$v_A = V \cos \omega t \tag{4.80}$$

$$v_B = V \cos(\omega t - \pi/2). \tag{4.81}$$

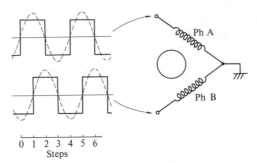

Fig. 4.12. When a stepping motor is driven from a bridge circuit, a two-phase square-wave voltage is applied to the phases. The fundamental component is indicated by the broken curves (after Ref. [4], reproduced by permission of Professor P. J. Lawrenson and Dr A. Hughes).

These equations can be solved to yield an expression for the torque produced by the motor at a steady speed corresponding to ω as

$$T = \frac{N_r n \Phi_M}{\sqrt{(r^2 + \omega^2 L)}} V \sin(\rho + \nu) - \frac{n^2 N_r \Phi_M^2 \omega r}{r^2 + \omega^2 L^2} \qquad (4.82)$$

where

$$\nu = \tan^{-1}(r/\omega L). \qquad (4.83)$$

In eqn (4.82), ρ is the rotor torque angle, and the maximum or pull-out torque T_p is obtained when ρ is such that $\rho + \nu = \pi/2$:

$$T_p = \frac{n N_r \Phi_M V}{\sqrt{(r^2 + \omega^2 L^2)}} - \frac{n^2 N_r \Phi_M^2 \omega r}{r^2 + \omega^2 L^2}. \qquad (4.84)$$

The significance of the two terms can be better appreciated, if the pull-out torque is expressed as a fraction of the peak static torque,† i.e. the torque at $\omega = 0$. From eqn (4.84) the maximum static torque is given by

$$T_M = n N_r \Phi_M V/r = n N_r \Phi_M I_M \qquad (4.85)$$

where I_M is the peak current in each phase. Then the normalized torque is

$$\frac{T_p}{T_M} = -\frac{1}{\sqrt{(1 + \omega^2 (L/r)^2)}} - k_p \frac{\omega(L/r)}{1 + \omega^2 (L/r)^2} \qquad (4.86)$$

where

$$k_p = \frac{n \Phi_M}{L(V/r)} = \frac{n \Phi_M}{L I_M}. \qquad (4.87)$$

---

† The peak static torque defined here is not exactly the same as the holding torque defined in Section 2.5.1.

4.4.2 Parameters governing torque–speed curve

The normalized torque expression in eqn (4.86) is simple and very informative. Firstly, it shows that if a motor has a given static torque, then its pull-out torque at a given speed or ω is determined by only two parameters; one is the stator time constant (L/r) and the second is k_p which is the ratio of flux linkage produced by the magnet to the self-flux linkage (LI) of the winding. The first term in eqn (4.86), being the inverse of the winding impedance, has the general shape shown by the chain curve (a) in Fig. 4.13, which decreases progressively at higher stepping rates. The second (negative) term is of the form shown by the broken curve (b). The resultant torque, shown by the solid curve (c), is the difference of the first and second term in eqn (4.87).

In Fig. 4.14 the effect of k_p on torque curves is shown for the winding time constant $L/r = 1$ ms. These curves show that the lower k_p, the higher the torque. It is said[4] that k_p lies in the range 0.2 to 0.6 in most hybrid motors. It is due to the braking effect expressed by the second (negative) term that the higher the k_p, the lower the torque, which corresponds to the fact that the higher the k_p, the better the damping.

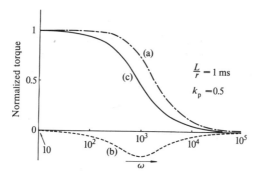

Fig. 4.13. Components of normalized pull-out torque curve.

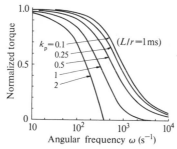

Fig. 4.14. Effect of k_p on normalized pull-out torque (after Ref. [4], reproduced by permission of Professor P. J. Lawrenson and Dr A. Hughes).

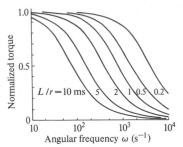

Fig. 4.15. Effects of L/r on pull-out torque (after Ref. [4], reproduced by permission of Professor P. J. Lawrenson and Dr A. Hughes).

The effects of varying the time constant by changing the series resistance (and changing the voltage to maintain the same current and static torque) are shown in Fig. 4.15 for a (typical) hybrid motor with $k_p = 0.25$. The addition of a forcing resistance increases the pull-out torque at all speeds. On the other hand, however, the heat loss in the resistance increases and the damping is degraded.

4.4.3 *Pull-in characteristics*

The pull-in torque versus speed curve represents the maximum frictional-load torque at which a stepping motor can start without failure of motion when a pulse train of the corresponding frequency is applied. In the analysis of this performance care must be taken to take account of the effect of inertia as well as the frictional load. A fundamental theory was presented by Lawrenson et al.,[5] but it is not reproduced here. To have good pull-in characteristics, several circuitry techniques have been invented and widely employed. These matters will be covered in Chapter 5.

4.5 Resonances and instabilities

The theory of pull-out torque presented in the previous section is developed on a simplified model with ideal assumptions. One of the important aspects this theory does not deal with is the resonance and instability which occur at certain stepping speeds. The resonance and instability are oscillatory phenomena which disturb the normal operation of the stepping motor. In some cases the magnitude of oscillation increases with time and eventually the motor loses synchronism. Actual pull-out torque curves have, in many cases, dips and islands as shown in Fig. 4.16. It is widely believed that the dips appear for the same reasons as instability.

The resonance and instabilities which appear in the pull-out curve may be classified into three categories: the low-frequency resonance, the

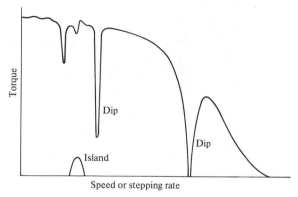

Fig. 4.16. Example of pull-out torque curve with dips and islands.

mid-range instability and the higher-range oscillation. Moreover, another kind of instability is present at starting. That is, a stepping motor can not start normally at some loads at certain stepping rates. These regions are indicated by dips and islands in the pull-in torque curve.

4.5.1 Low-frequency resonance

When a stepping motor is started at a very low speed and the pulse frequency is increased slowly, resonances first occur at subharmonics of the natural frequency which is ordinarily around 100 Hz. Then a major resonance will appear at and around the natural frequency. These oscillations occur below 200 Hz and are called low-frequency resonances. Figure 4.17 shows two examples of irregular motion due to resonance measured[6] on a 7.5° three-stack VR motor. The curves are measured traces of instantaneous rotor position against time: (a) is characteristic of the irregular types of motion occurring when operation is attempted near the resonant frequency (25 Hz in this case); and (b) shows the instantaneous transition to very stable double-speed reverse running (at 26 Hz in this case). The natural frequency of oscillation is around 100 Hz in this case. As the pulse rate is increased above the natural frequency, the magnitude of oscillation decreases and becomes stable.

In most practical situations the low-frequency resonances do not critically limit the performance of stepping motor systems since most motor/load combination can be instantaneously started at stepping rates well above the natural frequency.

4.5.2 Medium-range instability

When the stepping rate is increased into the range of 500 to 1500 Hz, stepping motors show troublesome behavioural features which are not of the resonant type above. They are due to inherent instability in the motor

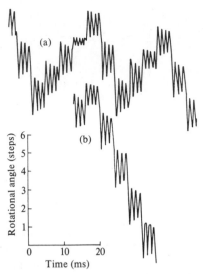

Fig. 4.17. Example of irregular motion due to resonance on a three-stack VR motor (after Ref. [6], reproduced by permission of the Institution of Electrical Engineers, and by courtesy of Professor P. J. Lawrenson). (a) Resonant behaviour at 25 Hz. (b) Onset of stable reverse running at double speed, 26 Hz.

or in the motor drive system. This kind of oscillation in stepping motors is known as the 'mid-frequency resonance' or 'mid-frequency instability'. The frequency of this oscillatory behaviour is quite different from the natural frequency, but 1/4 to 1/5 of the stepping rate. One of the most important problems which are encountered in applying a stepping motor to a system is how to avoid instability of this sort or overcome it. The characteristics of the mid-frequency instability are not very simple, and views are somehow different among papers. According to a paper[7] by Ward and Lawrenson instability of this sort has the following features.

(i) The oscillations may have one, or several frequency components. They are not simply related to the stepping rate, and are usually at a relatively low frequency, e.g. in the range 5–200 Hz.

(ii) With constant operating conditions, although synchronism can be lost suddenly, there is normally a slow build up of oscillations, over a period of many seconds or even minutes before failure finally occurs.

(iii) The various characteristics of the instability depend in a complex way on the type of drive and on the mode of operation; e.g. full- or half-step mode.

(iv) The size of torque dip and capacity to run to high speed are very sensitive to the degree of mechanical damping. Often with sufficient damping dips can be 'smoothed out' or the speed range is extended.

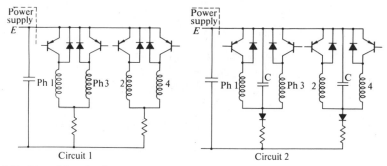

Fig. 4.18. Two drive circuits.

(v) Inertia is an important parameter; large inertias commonly amplify the problem.

G. Singh *et al.*[8],[9] studied the characters of the mid-range instability and methods of suppressing it by means of circuitry techniques. Some results are summarized here. The motor used for the studies is a 1.8° bifilar-wound hybrid motor rated for 4 A/phase operation. The holding torque is 2.1 N m with two-phase excitation at rated current, and the rotor inertia is 1.23×10^{-4} kg m^2. In reference [8] are shown four different drives used for measurement, and two of them are as shown in Fig. 4.18. Figures 4.19 and 4.20 show the plots of the amplitude of oscillation versus the stepping rate when the motor is operated without any load. The curves for the low-frequency resonance represent 1:1 (stepping rate) oscillations, related to the natural frequency. The natural frequencies measured from the single-step response are 143 Hz and 148 Hz for circuits 1 and 2, respectively. The loss of synchronism took place in the range 118–120 Hz with both circuits. The amplitude of the low-frequency

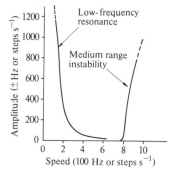

Fig. 4.19. Amplitudes of oscillations in the case of drive 1 (after Ref. [8], by courtesy of Dr G. Singh and Professor P. J. Lawrenson).

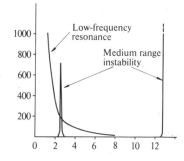

Fig. 4.20. Amplitudes of oscillations in the case of drive 2 (after Ref. [8], by courtesy of Dr G. Singh and Professor P. J. Lawrenson).

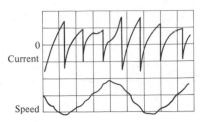

Fig. 4.21. Current waveform and speed when driven from drive 1 at 850 Hz. The period of speed fluctuation is about four times the pulse interval: vertical scale; 2 A/div, 556 Hz/div (after Ref. [8], by courtesy of Dr G. Singh and Professor P. J. Lawrenson).

resonance decreases steadily until it becomes negligible around 600 Hz. In this speed range the motor maintains synchronism in a very stable manner and can provide good load torque within its capability.

As is obvious from Figs. 4.19 and 4.20, the stepping rates at which the mid-frequency instability appears are different for drives 1 and 2. In drive 1 the instability starts at 760 Hz, but maintains synchronism up to 900 Hz. On the other hand in drive 2, a peak appears at 260 Hz, but this oscillation occurs in a very narrow frequency range and does not result in loss of synchronism. A large amplitude instability appears suddenly at 1275 Hz and synchronism fails. The frequency of those oscillations is in the range 1/5 to 1/4 of the stepping rate. The current and speed waveforms are shown in Fig. 4.21.

4.5.3 *Effects of system-parameter variations on instability*

The results so far presented have been for an unloaded motor. The effects of system-parameter variations on a loaded motor are reported[9] as follows:

(i) *Load torque.* In the case of drive 1 the occurrence of mid-range instability is delayed somewhat when the load is applied, while in the case of drive 2 the occurrence is advanced.

(ii) *Load inertia.* Presence of load inertia has an effect opposite to that of load torque. That is, in drive 1 the mid-range instability occurs earlier while in drive 2 it occurs later when load is present.

(iii) *Series resistance.* Higher drive voltages and higher series resistances improve the performance of both drives in the mid-range frequency.

(iv) *Condenser.* In the case of drive 2, the smaller the condenser C, the better the drive performs. But smaller values of C increase the peak collector-to-emitter voltage V_{CE} which must be withstood by the drive transistors in the OFF phases.

(v) *Dampers.* If the region of mid-range instability is to be crossed without loss of synchronism, the motor must be appropriately driven or an inertial damper must be used. While inertial dampers adversely affect response time, they may be useful in crossing mid-range instabilities if high responses are not required. When the use of a damper is not suitable, an appropriate drive and acceleration scheme are necessary. It is recommended that reference [9] be studied for this purpose.

4.5.4 *Higher-range oscillations*

If the frequency is increased further and the motor is successfully accelerated through the mid-range instability, the next region of instability occurs in the range 2500 to 4000 Hz But there are no reports which deal with this kind of oscillation.

4.5.5 *Theory of instability*

There are a number of theories of instability. But the theory presented by Hughes and Lawrenson[10] in 1979 is on the same lines as described so far in this chapter and informative on the fundamental mechanism of the instability and the effects of damping. Though attention is focused particularly on 1.8° hybrid motors in their theory, it is thought to be possible to derive similar results for VR motors because of the similarity of the dynamic equations of PM and VR motors.

(i) *Zero viscous damping.* When the damping coefficient D is zero, the general effect on the torque curve of the stability condition is shown in Fig. 4.22. The maximum torque follows the steady-state pull-out torque curve up to the point B, the breaking point, but then it diverges, following a new (lower) stability boundary to the point C. The shaded area in Fig. 4.22 indicates the unstable region. Note that with no viscous damping the theory does not predict a torque dip as such, but instead, the motor is

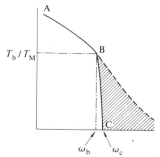

Fig. 4.22. Pull-out curve for zero damping (after Ref. [10], reproduced by permission of Professor P. J. Lawrenson and Dr A. Hughes).

unable to operate at any frequencies above ω_c. The analysis reveals that the break frequency is given by the simple formula

$$\omega_b = R/L \tag{4.88}$$

which corresponds to a stepping rate of $2R/\pi L$. A typical 1.8° hybrid motor and drive is expected to have a net time-constant in the range 1 to 0.5 ms, which gives the break points between 640 and 1280 Hz.

The steepness of the curve BC depends on the other parameters and r. For large total inertias (i.e. several times the motor inertia) and/or high values of r, the curve BC becomes almost vertical. And the lower the value of k_p, the steeper the line BC is. The torque at the break frequency, T_b, is given by

$$T_b/T_M = \frac{1}{\sqrt{2}} - \frac{k_p}{2} \tag{4.89}$$

where T_M is the maximum static torque given by eqn (4.85).

(ii) *Viscous damping and torque dips.* When $D \neq 0$, i.e. viscous damping is allowed for, a second stable region emerges at higher frequencies, which means that a dip may appear just as observed in practice (see Fig. 4.23). Relatively little damping is needed to introduce this high-speed stable region, and frequently air drag and friction will be sufficient. If enough damping is provided, the theory shows that the torque dip can be completely eliminated.

With viscous damping there are two break frequencies ω_{b1}, ω_{b2}. The break frequency ω_{b1} occurs at a slightly higher frequency than the break frequency ω_b for $D = 0$, while the width of the dip reduces as D increases. The shape and size of the dip are critically dependent on the damping, as shown in the enlarged view of the curves in Fig. 4.24, which show the

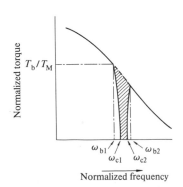

Fig. 4.23. Pull-out curve with damping (after Ref. [10], reproduced by permission of Professor P. J. Lawrenson and Dr A. Hughes).

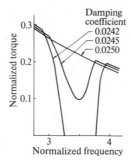

Fig. 4.24. Variation of torque dips with damping coefficient (after Ref. [10], reproduced by permission of Professor P. J. Lawrenson and Dr A. Hughes).

cases of $k_p = 0.4$ and $r/\omega_n L = 2$. The important condition for there to be enough damping to eliminate the dip is given by

$$D \gtrsim \frac{1}{8} \frac{J\omega_n^2 k_p}{R/L} . \tag{4.90}$$

4.6 Mechanical dampers

In applications of stepping motors it is undesirable that the normal operation is disturbed by resonance or instability. One of the counter-measures to damp out instabilities is to couple a mechanical damper to the motor shaft. In this section, types of mechanical dampers, their effects, and a fundamental theory of dampers are covered.

4.6.1 Constructions of dampers

There are several different kinds of damper, but the widely used ones are spring-friction inertial dampers (Fig. 4.25), magnet-friction inertial dampers (Fig. 4.26), and viscously-coupled inertial dampers or VCIDs (Fig. 4.27). The spring-friction damper consists of an inertial wheel which

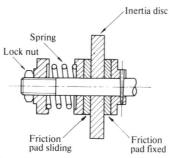

Fig. 4.25. A spring-friction inertial damper.

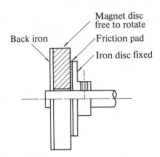

Fig. 4.26. A magnet-friction inertial damper.

is free to rotate on the shaft and mounted between two friction pads one of which is fixed to the shaft. Pressure between the disc and friction pad is maintained by means of a spring which can be adjusted to give optimum damping. The magnet damper consists of a free-wheeling disc of ferrite magnet and a steel disc directly coupled to the rotor. Between the two discs a friction pad is mounted, and a back iron is fixed on the magnet to provide flux path. Pressure is provided by magnetic attractive force between the discs and the friction pad. The VCID consists of an inertial wheel inside a cylindrical housing. The members can rotate freely relative to each other, but the annular space between them is small and filled with silicone fluid so that any relative movement leads to drag forces on both members. In operation with a stepping motor, the outer housing is rigidly coupled to the rotor shaft.

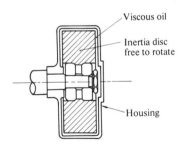

Fig. 4.27. A viscously-coupled inertial damper.

4.6.2 *Effects of dampers*

Let us assume that the rotor is revolving and oscillating. If the inertial wheel has a large inertia, it will tend to rotate at a constant speed and oscillating speed differences will appear between the rotor and inertial wheel. The viscousness or friction resists any speed difference and suppresses oscillation. For this reason, the bigger the inertia of the fly-wheel, the better the damping. On the other hand a big inertial wheel may degrade the acceleration capability and system efficiency. The optimization of the inertia of a damper is therefore an important problem.

A simple analytical explanation for the effect of the damper can be given by means of the transfer function of eqn (4.68) on p. 100, referring to the indicial response shown in Fig. 4.5 on p. 100. If the inertia of the flywheel is large, it may be supposed that the wheel is almost stationary in the early part of the indicial response, which means that the system's viscous coefficient D is large. It follows that the damping factor ζ is large and the indicial response becomes less oscillatory.

4.6.3 Theory of dampers

The above explanation is qualitative and assumes that the flywheel's inertia is very large Actually the flywheel's inertia must be suitably chosen. We are about to analyse the effect of the VCID of type (a) quantitatively. The torque developed in the motor, given by eqn (4.62), is written in the form

$$\tau_m = E_m(\theta_i - \theta_0) \tag{4.91}$$

where

$$E_m = K_T I_M N_r \tag{4.92}$$

$$\theta_i = \text{demanded rotor position}$$

$$\theta_0 = \text{actual rotor position.}$$

The equation of motion of the rotor is

$$(J_m + J_{di})\ddot{\theta}_0 = \tau_m - \tau_d \tag{4.93}$$

where J_m = inertia of the rotor
$\quad\;\; J_{di}$ = inertia of damper housing.
The viscous torque exerted on the housing, τ_d, is given by

$$\tau_d = D(\dot{\theta}_0 - \dot{\theta}_{d0}) \tag{4.94}$$

where θ_{d0} = position of the inertial flywheel. The equation of motion of the inertial wheel is

$$J_{d0}\ddot{\theta}_{d0} = \tau_d. \tag{4.95}$$

These equations define a third-order system and are solved by Lawrenson and Kingham.[11] The single-step operation with zero initial conditions yields the response of the rotor/damper housing combination in Laplace-transform terms as:

$$G(s) = \frac{\Theta_0}{\Theta_i} = \frac{E(K+s)}{s^3 + K(1+J)s^2 + Es + EK} \tag{4.96}$$

where

$$E = E_m/(J_m + J_{di}) \tag{4.97}$$

$$K = D/J_{d0} \tag{4.98}$$

$$J = \text{inertia ratio given by } J_{d0}/(J_m + J_{di}). \tag{4.99}$$

The indicial response is derived from eqn (4.96) as

$$\theta_0(t) = \theta_i\{1 - Ae^{-\alpha t} - Be^{-\beta t} \cos(\omega t + \phi)\} \tag{4.100}$$

where

$$A = J/4 \tag{4.101}$$

$$B = \frac{1}{4}\sqrt{\left(\frac{15J^4 + 72J^3 + 64J^2 + 128J + 256}{-J^2 + 8J + 16}\right)} \tag{4.102}$$

$$\omega = \frac{1}{4}\sqrt{\left(\frac{E(-J^2 + 8J + 16)}{J + 1}\right)} \tag{4.103}$$

$$\phi = \tan^{-1}\frac{4J(J+2)}{(J+4)\sqrt{(-J^2 + 8J + 16)}} \tag{4.104}$$

with

$$\alpha + 2\beta = K(J+1) \tag{4.105}$$

and β satisfying the following equation

$$8\beta^3 - 8K(J+1)\beta^2 + 2\{K^2(J+1)^2 + E\}\beta - EKJ = 0. \tag{4.106}$$

From these equations the fundamental idea required to optimize the damper can be derived. Readers who are interested in its details are recommended to read Reference [11]. But some of the important aspects will be reproduced here.

For most motor applications, it is desirable to achieve as rapid a decay in the oscillations as possible. This requires the maximum possible value of β and, if this can be achieved without a large value of α resulting (implying an unacceptably slow response build-up), then a good overall response can be achieved. Differentiating eqn (4.106) with respect to K, and taking $d\beta/dK = 0$ to obtain the maximum β, we have

$$K = \frac{8\beta^2(J+1) + EJ}{4\beta(J+1)^2} \tag{4.107}$$

and substituting into eqn (4.106)

$$\beta_{max} = \frac{J}{4}\sqrt{\left(\frac{E}{J+1}\right)}. \tag{4.108}$$

Corresponding to this, from eqn (4.105)

$$\alpha = \sqrt{\left(\frac{E}{J+1}\right)} \quad \text{for} \quad \beta_{max} \tag{4.109}$$

and from eqn (4.107)

$$K = \frac{J+2}{2(J+1)} \sqrt{\left(\frac{E}{J+1}\right)} \quad \text{for} \quad \beta_{max}. \tag{4.110}$$

Also, from eqns (4.108) and (4.109)

$$\frac{\beta}{\alpha} = \frac{J}{4} \tag{4.111}$$

which indicates that, as was hoped, α need not become large.

It can be shown that the response is non-oscillatory if the inertia ratio J satisfies the inequality

$$J^2 - 8J - 16 \geqq 0 \tag{4.112}$$

or

$$J \geqq 4(1 + \sqrt{2}) \simeq 9.66. \tag{4.113}$$

This is, however, impractical since such a large inertia ratio degrades acceleration rates.

A practical optimum can be expected when $\alpha = \beta$ and $J = 4$. Figure 4.28 shows the graphs of α, β, and ω as function of J. As the inertia ratio J increases the value of β increases, while α decreases. It is an idea of optimization of the system to have $\alpha = \beta$ which occurs in the region where J is not so big. Lawrenson and Kingham.[11] confirmed that $J = 4$ is also a reasonable optimum condition for a model of an actual non-linear torque vs. displacement characteristic.

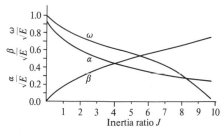

Fig. 4.28. Variations of α, β, and ω with inertia ratio J (after Ref. [11], redrawn by permission of the Institution of Electrical Engineers and by courtesy of Professor P. J. Lawrenson).

The optimum value of D, for $J = 4$ in the linear theory, is

$$D = \frac{J+2}{2(J+1)} \sqrt{\left(\frac{J}{J+1}\right)} \sqrt{(E_m J_{d0})} = 0.54\sqrt{(E_m J_{d0})}. \qquad (4.114)$$

References for Chapter 4

[1] Hughes, A. and Lawrenson, P. J. (1975). Electromagnetic damping in stepping motors. *Proc. IEE* **122,** (8) 819–24.
[2] Hughes, A. and Lawrenson, P. J. (1978). U.K. Patent No. 1,523,348.
[3] Kenjo, T. and Niimura, Y. (1979). *Fundamentals and applications of stepping motors.* (In Japanese.) p. 111. Sogo Electronics Publishing Co., Ltd., Tokyo.
[4] Hughes, A., Lawrenson, P. J., and Davies, T. S. (1976). Factors determining high-speed torque in hybrid motors. *Proc. International conference on stepping motors and devices.* University of Leeds, pp. 150–7.
[5] Lawrenson, P. J., Hughes, A., and Acarnley, P. P. (1976). Starting/stopping rates of stepping motors: Improvement and prediction. *Proc. International conference on stepping motors and systems.* University of Leeds, pp. 54–60.
[6] Lawrenson, P. J. and Kingham, I. E. (1977). Resonance effects in stepping motors. *Proc. IEE* **124,** (5), 445–8.
[7] Ward, P. A. and Lawrenson, P. J. (1977). Backlash, resonance and instability in stepping motors. *Proc. Sixth annual symposium on Incremental motion control systems and devices.* Department of Electrical Engineering, University of Illinois, pp. 73–83.
[8] Singh, G., Leenhouts, A. C., and Mosel, E. F. (1976) Electromagnetic resonance in permanent-magnet step motor drive system. *Proc. International conference on stepping motors and systems.* University of Leeds, pp. 115–24.
[9] Leenhouts, A. C. and Singh, G. (1977). An active stabilization technique for open loop permanent-magnet step motor drive system. *Proc. Sixth annual symposium on Incremental motion control systems and devices.* Department of Electrical Engineering, University of Illinois, pp. 19–24.
[10] Hughes, A. and Lawrenson, P. J. (1979). Simple theoretical stability criteria for 1.8° hybrid motors. *Proc. International conference on stepping motors and systems.* University of Leeds, pp. 127–35.
[11] Lawrenson, P. J. and Kingham, I. E. (1975). Viscously coupled inertial damping of stepping motors. *Proc. IEE* **122,** (10), 1137–40.

5. Drive system and circuitry for open-loop control of stepping motors

One of the most important problems in stepping-motor application is the drive system. The drive systems of stepping motors are classified into open-loop and closed-loop schemes. This chapter discusses the open-loop drive system.

5.1 Drive system

A simple drive system for a stepping motor is represented by the block diagram in Fig. 5.1, the number of phases being four in this example. The block diagram is divided into two portions for convenience of explanation. Figure 5.1(a) represents the portion from logic sequencer to motor. When a step-command pulse is applied to the logic sequencer, the states of the output terminals are changed to control the motor driver so as to rotate the motor a step angle in the desired direction. The rotational direction is determined by the logic state at the direction input, e.g. the H level for the CW and the L for the CCW direction. In some applications the logic sequencer is unidirectional, having no direction-signal terminal. If one increment of movement is performed by one step, the block diagram of Fig. 5.1(a) represents the whole system. But, when an increment is performed by two or more steps, another stage to produce a proper train of pulses is needed to put before the logic sequencer, and this is represented in Fig. 5.1(b). This logic circuit is termed the 'input

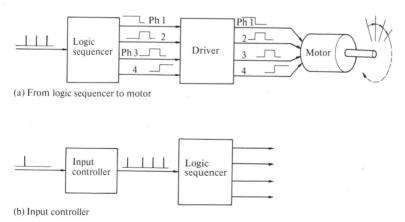

(a) From logic sequencer to motor

(b) Input controller

Fig. 5.1. Block diagram of the drive system of a stepping motor.

controller' in this book. In sophisticated applications the function of input controller is carried out by an intelligent electronic device like a micro-processor which generates a pulse train to speed up, slew, and slow down the motor in the most efficient and reliable manner. In this chapter the details of pulse sequencers are first discussed, and there then follows discussion of power drivers and input controllers. Lastly an example of microprocessor application to the open-loop control will be presented.

5.2 Logic sequencers

The logic sequencer is a logic circuit which controls the excitation of windings sequentially, responding to step-command pulses. A logic se-quencer is usually composed of a shift register and logic gates such as NANDs, NORs, etc. Nowadays, shift-register IC chips for use are univer-sally available. But one can assemble a logic sequencer for a particular purpose by a proper combination of J–K flip–flop (JK-FF) IC chips and logic-gate chips. The fundamental functions of the gates and JK-FF are summarized in Table 5.1. Instead of assembling a sequencer with discrete IC chips or/and a shift register, purpose-built logic sequencers designed for stepping motors are available.

In this section a number of types of logic sequencers assembled with TTL ICs will be first presented, and then a universal sequencer of CMOS type is discussed.

5.2.1 *Two-phase-on excitation for a four-phase motor*

A type of simple sequencer can be built with only two JK-FFs, as shown in Fig. 5.2 for the unidirectional case. The truth tables of the logic sequence are given in the same figure. The correspondence between the output terminals of the sequencer and the phase windings to be controlled is as follows.

$$Q1 \text{ —————— } Ph1$$
$$\overline{Q1} \text{ —————— } Ph2$$
$$Q2 \text{ —————— } Ph3$$
$$\overline{Q2} \text{ —————— } Ph4$$

That is, if Q1 is on the H level the winding Ph1 is excited, and if Q1 is on the L level Ph1 is not excited. As compared in the two tables, circuits (a) and (b) are opposite in the sequence of excitation, the direction of circuit (a) being defined as CW (= clockwise) and that of (b) as CCW (= counter-clockwise). To reverse the rotational direction, the connections of the sequencer must be interchanged between (a) and (b). The direction-switching circuits shown in Fig. 5.3 may be used for this purpose, the

Table 5.1. Logic gates and their function.

AND	A·B=C	Input A	Input B	Output C
		1	1	1
		1	0	0
		0	1	0
		0	0	0

NAND	$\overline{A \cdot B} = C$	Input A	Input B	Output C
		1	1	0
		1	0	1
		0	1	1
		0	0	1

OR	A+B=C	Input A	Input B	Output C
		1	1	1
		1	0	1
		0	1	1
		0	0	0

NOR	$\overline{A + B} = C$	Input A	Input B	Output C
		1	1	0
		1	0	0
		0	1	0
		0	0	1

NOT	$\overline{A} = B$	Input A	Output B
		1	0
		0	1

JK-FF	Clear	Input J	K	Output Q_{t+1}	$\overline{Q}_{t+1}$	
		1	1	$\overline{Q}_t$	Q_t	
		1	0	1	0	set
		0	1	0	1	reset
		0	0	Q_t	$\overline{Q}_t$	

The function of the table is effected when the clear terminal is on level H. If the clear terminal is on level L, output Q will be on L and $\overline{Q}$ on H.

essential function being in the combination of three NAND gates or two AND gates and a NOR gate. In circuit (a), if the direction-command signal is on the H level the same level as at input terminal A appears at the output terminal C. Conversely, if the direction command is on the L level, the signal at C is the same as that at the input terminal B. In circuit (b), $C = \overline{A}$ for the H-level direction command, and $C = \overline{B}$ for the direction

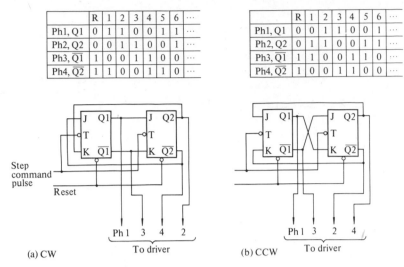

	R	1	2	3	4	5	6	
Ph1, Q1	0	1	1	0	0	1	1	...
Ph2, Q2	0	0	1	1	0	0	1	...
Ph3, $\overline{Q1}$	1	0	0	1	1	0	0	...
Ph4, $\overline{Q2}$	1	1	0	0	1	1	0	...

	R	1	2	3	4	5	6	
Ph1, Q1	0	0	1	1	0	0	1	...
Ph2, Q2	0	1	1	0	0	1	1	...
Ph3, $\overline{Q1}$	1	1	0	0	1	1	0	...
Ph4, $\overline{Q2}$	1	0	0	1	1	0	0	...

(a) CW   To driver   (b) CCW   To driver

Fig. 5.2. A unidirectional logic sequencer for two-phase-on operation of a four-phase motor.

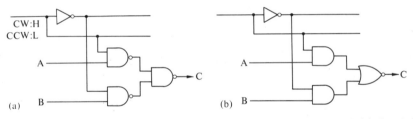

Fig. 5.3. Logic selectors used for commanding rotational direction; in circuit (a) C = A for the H-level command and C = B for the L-level command, while in (b), C = $\overline{A}$ for the H command and C = $\overline{B}$ for the L command.

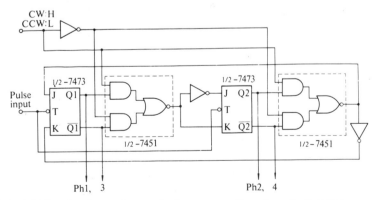

Fig. 5.4. A bidirectional two-phase-on logic sequencer for a four-phase motor.

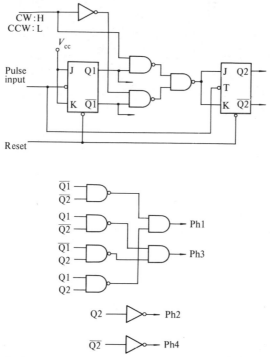

Fig. 5.5. Another logic sequencer for two-phase-on operation of a four-phase motor.

command of L level. Figure 5.4 shows a bidirectional sequencer for the two-phase-on excitation for a four-phase stepping motor which incorporates two direction-switching portions. Another example is given in Fig. 5.5.

5.2.2 *Single-phase-on sequencers for four-phase motor*

A single-phase-on sequencer is realized by adding four AND gates to the output terminals of the two-phase-on sequencer, as shown in Figs. 5.6 and 5.7. The truth tables of the logic sequence are shown in Table 5.2 (p. 162) for both CW and CCW directions.

In the circuits in this book, like most cases, the changes in the output states take place at the build-down of the step-command pulse. The circuit which is put in front of the sequencer to eliminate noise and reform the waveform is shown later in Fig. 5.36 (p. 145).

5.2.3 *One-phase-on sequencer for three-phase motors*

Unidirection sequencers for the one-phase-on drive of a three-phase motor are composed with a shift register and three AND gates, as shown

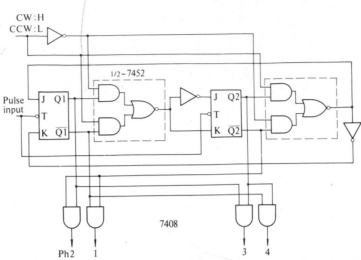

Fig. 5.6. A bidirectional logic sequencer for single-phase-on operation of a four-phase motor.

		R	1	2	3	4	5	6	
	Q_1	0	1	0	1	0	1	0	...
	$\overline{Q}_1$	1	0	1	0	1	0	1	...
	Q_2	0	0	1	1	0	0	1	...
	$\overline{Q}_2$	1	1	0	0	1	1	0	...
Ph1	$\overline{Q}_1 \cdot \overline{Q}_2$	1	0	0	0	1	0	0	...
Ph2	$Q_1 \cdot \overline{Q}_2$	0	1	0	0	0	1	0	...
Ph3	$\overline{Q}_1 \cdot Q_2$	0	0	1	0	0	0	1	...
Ph4	$Q_1 \cdot Q_2$	0	0	0	1	0	0	0	...

CW

		R	1	2	3	4	5	6	
	Q_1	0	1	0	1	0	1	0	...
	$\overline{Q}_1$	1	0	1	0	1	0	1	...
	Q_2	0	1	1	0	0	1	1	...
	$\overline{Q}_2$	1	0	0	1	1	0	0	...
Ph1	$\overline{Q}_1 \cdot \overline{Q}_2$	1	0	0	0	1	0	0	...
Ph2	$Q_1 \cdot \overline{Q}_2$	0	0	0	1	0	0	0	...
Ph3	$\overline{Q}_1 \cdot Q_2$	0	0	1	0	0	0	0	...
Ph4	$Q_1 \cdot Q_2$	0	1	0	0	0	1	0	...

CCW

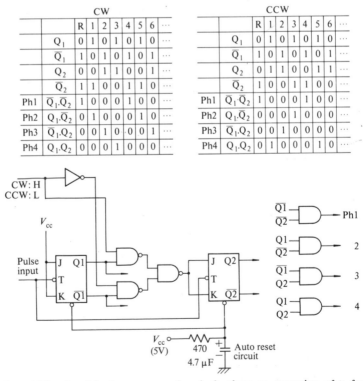

Fig. 5.7. A bidirectional logic sequencer for single-phase-on operation of a four-phase motor.

CW

	R	1	2	3	4	5	
Q_1	0	1	0	0	1	0	...
$\overline{Q}_1$	1	0	1	1	0	1	...
Q_2	0	0	1	0	0	1	...
$\overline{Q}_2$	1	1	0	1	1	0	...
Ph1 $\overline{Q}_1.\overline{Q}_2$	1	0	0	1	0	0	...
Ph2 $Q_1.\overline{Q}_2$	0	1	0	0	1	0	...
Ph3 $\overline{Q}_1.Q_2$	0	0	1	0	0	1	...

CCW

	R	1	2	3	4	5	
Q_1	0	0	1	0	0	1	...
$\overline{Q}_1$	1	1	0	1	1	0	...
Q_2	0	1	0	0	1	0	...
$\overline{Q}_2$	1	0	1	1	0	1	...
Ph1 $\overline{Q}_1.\overline{Q}_2$	1	0	0	1	0	0	...
Ph2 $Q_1.\overline{Q}_2$	0	0	1	0	0	1	...
Ph3 $\overline{Q}_1.Q_2$	0	1	0	0	1	0	...

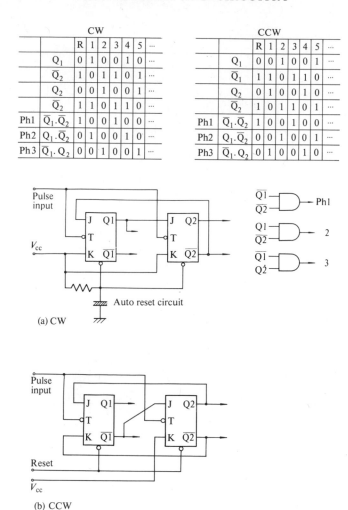

Fig. 5.8. Unidirectional pulse sequencer for three-phase motors.

in Fig. 5.8. Circuit (a) is for the CW direction and (b) for the CCW direction. The truth tables of the logic sequence are shown in the same figure. The bidirectional sequencer is shown in Fig. 5.9 in which two direction-switching circuits are used to switch the connections between two JK-FFs.

5.2.4 Two-phase-on sequencer for three-phase motors

This logic sequencer is obtained by using three NAND gates instead of AND gates. See also Fig. 5.9.

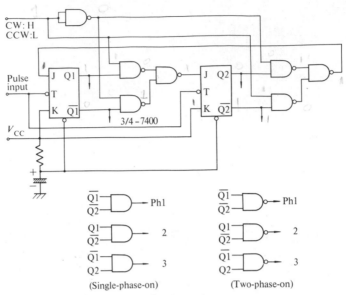

Fig. 5.9. Bidirectional logic sequencers for three-phase motors.

		R	1	2	3	4	5	6	7
					CW				
Ph1,	$\overline{Q1}$	1	1	0	0	0	1	1	1
Ph2,	Q2	0	0	0	1	1	1	0	0
Ph3,	Q3	0	1	1	1	0	0	0	1

		R	1	2	3	4	5	6	7
					CCW				
Ph1,	$\overline{Q1}$	1	1	0	0	0	1	1	1
Ph2,	Q2	0	1	1	1	0	0	0	1
Ph3,	Q3	0	0	0	1	1	1	0	0

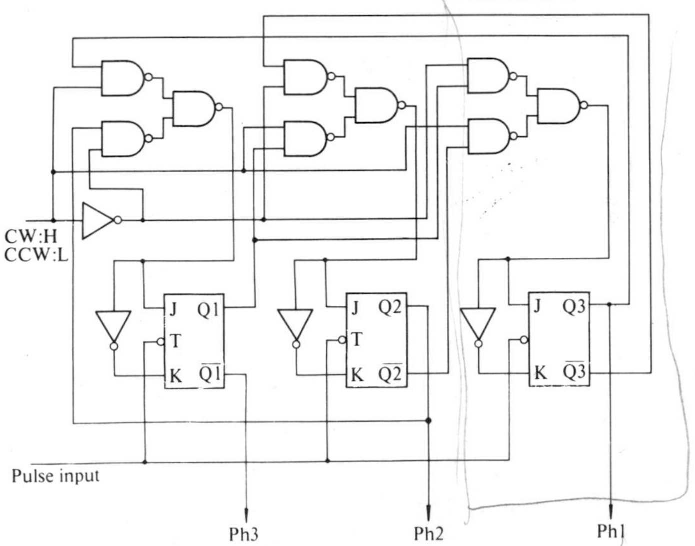

Fig. 5.10. Logic sequencer for half-step operation of a three-phase motor.

5.2.5 Half-step sequencer for three-phase motors

An example is shown in Fig. 5.10 which uses three JK-FFs.

5.2.6 Two-phase-on sequencer for bifilar-wound three-phase motor of VR type

The sequence required for two-phase-on drive of a bifilar-wound three-phase motor is shown in Table 2.4 on p. 51. This is obtained by putting six AND or NOR gates after the output terminals of the circuit in Fig. 5.10, the connections being indicated in Fig. 5.11.

5.2.7 Two forms of bidirectional sequencer

The pulse sequencers of bidirectional type so far explained have the step-command terminal and direction terminal on the input side. When this type of sequencer is seen as a black box, it will be such as shown in Fig. 5.12(a). There is another type of sequencer which has an input terminal for CW steps and for CCW steps, the black-box illustration is as Fig. 5.12(b).

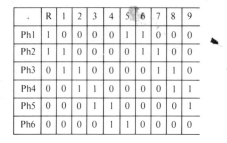

	R	1	2	3	4	5	6	7	8	9
Ph1	1	0	0	0	0	1	1	0	0	0
Ph2	1	1	0	0	0	0	1	1	0	0
Ph3	0	1	1	0	0	0	0	1	1	0
Ph4	0	0	1	1	0	0	0	0	1	1
Ph5	0	0	0	1	1	0	0	0	0	1
Ph6	0	0	0	0	1	1	0	0	0	0

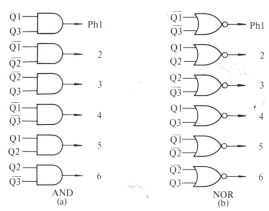

Fig. 5.11. For obtaining logic sequencer for bifilar wound three-phase motor, the above gates are added to the outputs of the circuit in Fig. 5.10.

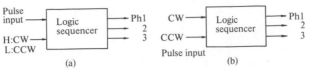

Fig. 5.12. Two forms of bidirectional pulse sequencer.

5.2.8 *Sequencer for bridge drive of four-phase hybrid motors*

The arrangement of switching devices for the bridge drive of a hybrid stepping motor is shown in Figs. 2.62 and 2.63. For the scheme in Fig. 2.63 using eight power transistors the switching devices S1 and S4 must work simultaneously. Likewise the pairs S2 and S3, S5 and S8, and S6 and S7 work simultaneously. The logic sequencer in Fig. 5.4 can be utilized to drive these switching devices by the following connections:

   output of $\overline{Q2}$ is for S1 and S4
   output of Q2 is for S2 and S3
   output of Q1 is for S5 and S8
   output of $\overline{Q1}$ is for S6 and S7.

For the driver scheme of Fig. 2.63 (p. 57) the connections can be as follows:

   output of $\overline{Q2}$ is for S1
   output of Q2 is for S2
   output of Q1 is for S3
   output of $\overline{Q1}$ is for S4.

5.2.9 *Universal sequencer MSI*

Instead of assembling a logic sequencer with discrete IC chips to operate a particular driver in a particular mode, one may use some MSI sequencer

Fig. 5.13. Sanyo's CMOS monolithic logic sequencer for universal use.

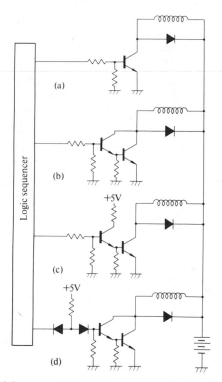

Fig. 5.14. Examples of the connection between a sequencer and a driver.

designed to serve for several different operational modes of the stepping motor. For example, Sanyo's PMM8713 shown in Fig. 5.13 is a CMOS monolithic MSI designed to control either a three- or four-phase stepping motor in any of the single-phase-on, two-phase-on, or half-step excitations. The output signals from these kinds of universal logic sequencer can directly control the bases of Darlington power transistors connected as in the circuit in Fig. 5.14(b)–(d).

5.3 Motor driver

5.3.1 *Connection of sequencer and driver*

Output signals of a logic sequencer are transmitted to the input terminals of a power driver, by which the turning on/off of the motor windings is governed. The power driver may be called a 'motor driver' or simply a 'driver'. The simplest method of connection is the direct connection such as that shown in Fig. 5.14(a) and (b). But, if the output currents from the sequencer are not enough to drive the power transistors, it is necessary to put a buffer for current amplification between the two stages, as shown in Fig. 5.14(c) and (d).

5.3.2 *Problems with drivers*

A winding on a stepping motor is inductive and appears as a combination of inductance and resistance in series. In addition, as a motor revolves, a counter e.m.f. is produced in the winding. The equivalent circuit to a winding is, hence, such as that shown in Fig. 5.15. On designing a power driver, one must take into account necessary factors and behaviour of this kind of circuit. Firstly, the worst-case conditions of the stepping motor, power transistors, and supply voltages must be considered. The motor parameters vary due to manufacturing tolerances and operating conditions. Since stepping motors are designed to deliver the highest power from the smallest size, the case temperature can be as high as about 100 °C, and the winding resistance therefore increases by 20 to 25 per cent.

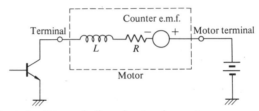

Fig. 5.15. Equivalent circuit to a winding of a stepping motor.

5.3.3 *Suppressors*

When the transistor in Fig. 5.15 is turned off, a high voltage builds up due to $L(di/dt)$, and this voltage may damage the transistor. There are several methods of suppressing this spike voltage and protecting the transistor, as shown in the following:

(1) *Diode suppressor.* If a diode is put in parallel with the winding in the polarity as shown in Fig. 5.16 a circulating current will flow after the transistor is turned off, and the current will decay with time. In this scheme, no big change in current appears at turn-off, and the collector potential is the supply potential E plus the forward potential of the diode. This method is very simple, but a drawback is that the circulating current lasts for a considerable length of time and it produces a braking torque.

(2) *Diode/resistor suppressor.* A resistor is connected in series with the diode as in Fig. 5.17 to damp quickly the circulating current. The voltage V_{CE} applied to the collector at turn-off in this scheme is

$$V_{CE} = E + IR_S + V_{DF}$$

Fig. 5.16. Diode suppressor.   **Fig. 5.17.** Diode-resistor suppressor.

where E = supply potential

     I = current just before turning-off

     R_S = resistance of suppressor resistor

     V_{DF} = forward potential of diode.

The higher the resistance R_S, the quicker the current decays after turn-off, but the higher the collector potential. Therefore, a higher maximum voltage rating is required for fast decay.

(3) *Zener diode suppressor.* Zener diodes are often used to connect in series with the ordinary diode as shown in Fig. 5.18. Compared with the preceding two cases, in this scheme the current decays more quickly after turn-off (see Fig. 5.19). In addition to this, it is a merit of this method that the potential applied to the collector is the supply potential plus the zener potential, independent of the current. This makes the determination of the rating of the maximum collector potential easy. Figure 5.20 is a driver circuit with zener diode suppression for a four-phase motor. Resistors $R1$ and $R2$ are for quick build-up of exciting current, which will be explained in 5.3.4.

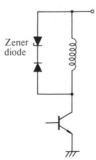

Fig. 5.18. Zener diode suppressor.

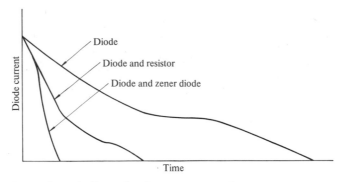

Fig. 5.19. Comparison of effects of various suppressor schemes.

(4) *Condenser suppressor.* This scheme is often employed for bifilar-wound motors. An explanation is given for the circuit shown in Fig. 5.21, which is for a four-phase motor. A condenser is put between Ph1 and Ph3 and between Ph2 and Ph4. These condensers serve a twofold purpose. Firstly, when a transistor is turned off, the condenser connected to it via a diode absorbs the decaying current from the winding to protect the transistor. Let us see the situation just after the transistor Tr1 is turned off in the one-phase-on mode. Either Tr2 or Tr4 is turned on, but Tr3 is still in the turned-off state. Since the windings of Ph1 and Ph3 are wound in the bifilar fashion, a transient current will circulate as indicated by the dotted loop in the figure. If Tr3 is turned on when the transient circulating current becomes zero and the charge stored in the condenser becomes maximum, a positive current can easily flow through Ph1 winding. By this resonance mechanism, currents are used efficiently in this scheme. This feature remains in the two-phase-on mode, too. The condenser suppressor is suited to drives in which the stepping rate is limited in a narrow region.

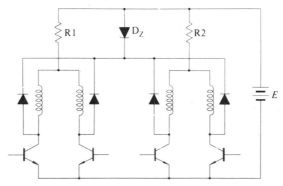

Fig. 5.20. Example of four-phase driver with zener diode suppressor.

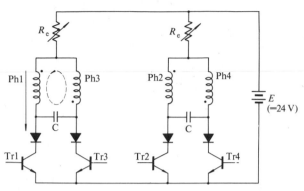

Fig. 5.21. Four-phase driver with condenser suppressor. External resistors denoted by R_e are adjusted so that the current is at rating.

Another utility of condensers is as an electrical damper. As stated in Section 4.2, a method of damping rotor oscillations is to provide a mechanism to convert kinetic energy into Joule heating. If a rotor having a permanent magnet oscillates, an alternating e.m.f. is generated in the winding. However, if a current path is not provided or a high resistance is connected, no current will be caused by this e.m.f. When a condenser is connected between phases, an oscillatory current will flow in the closed loop shown in Fig. 5.21, and Joule heat is generated in the windings, which means that the condenser works as an electrical damper. Variations of pull-out curve with capacitance are shown in Fig. 5.22, which was

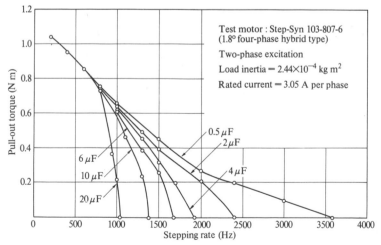

Fig. 5.22. Variation of pull-out torque curve with condenser connected between phases, measured on the circuit of Fig. 5.22 (from Ref. 2).

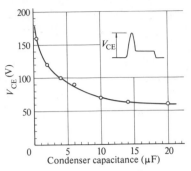

Fig. 5.23. Potential applied to the turned-off transistor as varied with capacitance, at maximum pull-out rate (from Ref. 2).

prepared using a hybrid motor.[2] The smaller the capacitance, the more the pull-out torque at higher stepping rates, which is due to quick decay of current after turning-off. Instead, the maximum potential applied to the collector after turning off becomes higher with decreasing capacitance, as shown in Fig. 5.23.

5.3.4 *Improvement of current build-up*

When a transistor is turned on to excite a phase, the power supply must overcome the effect of winding inductance before driving at the rated current, since the inductance has a tendency to oppose the current build-up in this case. As switching frequency increases, the build-up time to a cycle becomes large and it results in decreased torque and slow response. There are several methods of shortening the build-up time and improving the torque characteristics at high speeds, as will be outlined here.

(1) *Series resistance.* The least expensive way is to add a resistor in series with a winding as shown in Fig. 5.24. The power supply potential E is

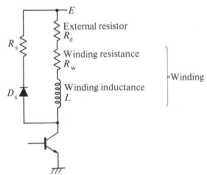

Fig. 5.24. Improving build-up by putting R_e in series with the winding and raising supply potential E.

selected to drive the rated current through windings under steady-state conditions. The time constant of the circuit is decreased from L/R_w to $L/(R_e + R_w)$.

Though the series resistance is the simplest method, it is disadvantageous in that much power is dissipated in the series resistors. If the winding resistance of a four-phase motor is $1.5 \,\Omega$ and the rated current is 4 A, and it is required to drive from a 24 V supply, then the resistance to be added is $4.5 \,\Omega$ in each phase. The power loss is approximately $4.5 \times 4^2 = 72$ W in the one-phase-on mode. It will be doubled in the two-phase-on drive.

(2) *Dual voltage.* To reduce the power dissipation in the driver and increase the performance of a stepping motor, the dual-voltage driver is used. The scheme for one phase is shown in Fig. 5.25. When a step-command pulse is given to the sequencer, a high-level signal will be put out from one of the output terminals, to excite a phase winding. On this signal both Tr1 and Tr2 are turned on, and the higher voltage E_H will be applied to the winding. The diode D1 is now reverse-biased to isolate the lower voltage supply from the higher voltage supply. The current builds up quickly due to the higher voltage E_H. The time constant of the monostable multivibrator is selected so that transistor Tr1 is turned off when the winding current exceeds the rated current by a little. After the higher voltage source is cut off, the diode is forward-biased and the winding current is supplied from the lower voltage supply. A typical current waveform is shown in Fig. 5.26.

When the dual-voltage method is employed for the two-phase-on drive of a four-phase motor, the circuit scheme will be such as that shown in Fig. 5.27. Two transistors are used for switching the higher voltage in

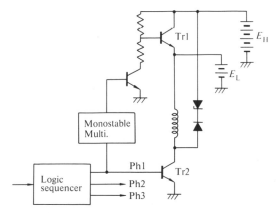

Fig. 5.25. Improvement of current build-up by means of dual voltage drive.

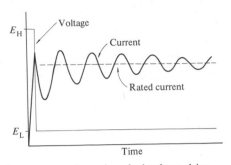

Fig. 5.26. Voltage and current waveforms in a dual voltage drive.

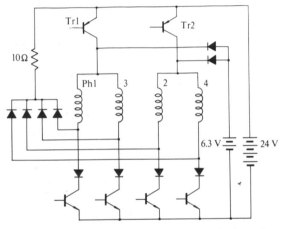

Fig. 5.27. A dual-voltage driver for the two-phase-on drive of a four-phase motor.

order that the winding once fed from the higher voltage is not again excited by it when a new phase is turned on.

In the dual voltage scheme, as the stepping rate is increased, the high voltage is turned on for a greater percentage of the time.

5.3.5 *Bridge drive of a four- or two-phase motor*

Motor efficiency is higher if all the windings are always excited in such a manner as to produce effective torque, compared with the case in which each winding is excited only for a limited percentage of the time. This is particularly true with small motors. In driving a four-phase hybrid motor, the bipolar drive is an excellent drive for this reason, since four windings are always excited. In comparison with the unipolar drive, in which the current flows in one direction in the winding, 20 to 35 per cent improvement in torque is possible.

There are two basic types of bridge circuit for driving a motor in the

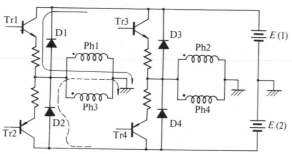

Fig. 5.28. A bridge circuit for the bipolar drive.

bipolar scheme. One is shown in Fig. 5.28, which uses two power supplies. The properties and functions of this scheme are summarized as follows:

1. Connection of windings. If the motor has eight separate terminals, the windings are connected as in the figure. An alternating current flows through each winding, Ph1 and Ph3 always having the same polarity. The motor, hence, is thought of as a two-phase motor.

2. Counter-measures to transistor damage. If transistor Tr1 is ON then Tr2 is OFF and vice versa. But there is a possibility that a transistor is turned on while the other is still conducting. So it is necessary to incorporate delay circuitry to prevent the two transistors from being in the ON state at the same time. Alternatively resistors may be added as in the figure so that any slight overlap of device conduction is not harmful.

3. Diodes to suppress spike voltages. The diodes connected in parallel with the power transistors are for suppressing the spike voltage which occurs when any transistor is turned off. This function of diodes is somewhat different from that in the unipolar drive. If Tr1 is conducting, the current path will be as shown by the light solid curve in Fig. 5.28. Just after Tr1 is turned off and Tr2 is turned on, the current in the winding will not yet have reversed, and it will circulate through D2 and the battery E(2) as shown by the light dotted curve. Now it may be said that the power supply E(2) provides a current from the negative terminal, or in other words, the battery is now being charged. In terms of energy, magnetic field energy in the windings is fed back to the power supply. Since in the unipolar drive the magnetic energy of this kind is consumed in windings, diodes, external resistors, and zener diodes, the bipolar drive is advantageous over it in this respect.

4. Reverse of current direction. In the unipolar drive, some form of suppressor circuit is added to damp the winding current after the turning-off of the corresponding transistor. On the other hand in the bipolar drive a voltage is applied to the winding to reverse the current. After the

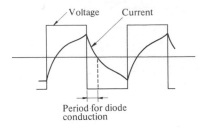

Fig. 5.29. Voltage and current waveforms in the bipolar drive.

current flowing along the dotted path becomes zero, another current will build up through Tr2 in the reverse direction. The relation between the voltage across the winding and the current through it is such as in Fig. 5.29. The current waveform is different from an exponential wave, but slightly concave due to the counter e.m.f. generated by the rotor motion.

Another scheme for a bipolar drive is shown in Fig. 5.30; this uses a single power supply but four transistors for each phase. The resistor connected in parallel to the windings is for quick build-up of excitation current. Time-delay circuits or resistors to prevent transistors from being damaged due to overlap in conduction are also needed in this case.

5.3.6 *Driver for three-phase bifilar-wound VR motor*

A driver[3] for a three-phase VR motor with bifilar windings is shown in Fig. 5.31. This driver is equipped with several excellent features as described in the following.

1. Diodes D1 to D6 are similar to those in bridge drivers in their function. In the R (= reset) state in the table in Fig. 5.31, the windings of Ph1 and Ph2 are excited. Then, when a step-command pulse comes, Ph1 is turned off and Ph3 is switched on, but Ph2 remains excited. In the transition period, the magnetic energy in Ph1 must be released in some form. Since the windings of Ph1 and Ph4 are wound in the bifilar fashion, to make magnetic coupling tight, a current will flow upward through the

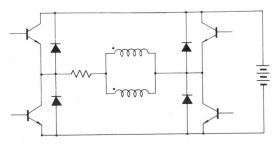

Fig. 5.30. Another bridge driver.

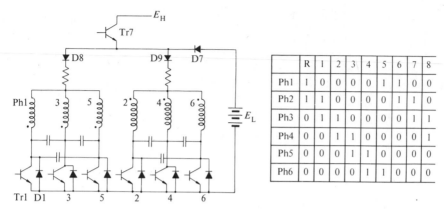

	R	1	2	3	4	5	6	7	8
Ph1	1	0	0	0	0	0	1	1	0	0
Ph2	1	1	0	0	0	0	0	1	1	0
Ph3	0	1	1	0	0	0	0	0	1	1
Ph4	0	0	1	1	0	0	0	0	0	1
Ph5	0	0	0	1	1	0	0	0	0
Ph6	0	0	0	0	1	1	0	0	0

Fig. 5.31. Dual-voltage driver for a three-phase VR motor with bifilar windings; Ph1 and Ph2, Ph3 and Ph6, and Ph5 and Ph2 are each wound in bifilar form.

windings of Ph4 and diode D4. This current circulates through the windings of Ph2 and transistor Tr2.

2. Condensers connecting windings provide a current path for the transient current in the early part of the transient phenomenon. For example, just after Tr1 is turned off and Tr3 is turned on, a transient current will flow from Ph1 to Tr3 via the condenser connecting Ph1 and Ph3.

3. Diodes D8 and D9 are for isolating the individual groups of windings from each other allowing only mutual coupling effects.

4. Diode D7 isolates the low-voltage supply E_L from the high voltage supply E_H when Tr7 is conducting.

5. Electric damping is effective due to the condensers and bifilar windings. That is, if the rotor motion is oscillatory, the back e.m.f. induced in each phase winding has fluctuating components which act with the mutual inductance of the bifilar windings to produce oscillating currents through the condensers. This is a mechanism by which oscillating kinetic energy is damped out and converted into Joule heating.

5.3.7 Pulse-width modulation drive

The pulse-width modulation (= PWM) driver is an excellent driver which offers good current build-up with low loss. The basic function of a PWM driver is illustrated in Fig. 5.32. Here the inductive load put in the square drawn by broken lines represents an ordinary driver (see Fig. 5.33). The voltage at the load current pick-up is compared with the reference voltage by means of an operational amplifier with a high gain. The reference voltage is the superposition of a high-frequency triangle or sinusoidal component and a DC component with which the current pick-up voltage

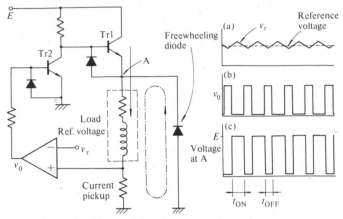

Fig. 5.32. PWM (= pulse width modulated) circuit and waveforms.

is to be compared. If the DC component of the reference signal and the pick-up voltage are almost the same, the waveform at the output terminal of the amplifier will be a square wave as shown in Fig. 5.32(b). Since the gain of the amplifier is high, the output voltage oscillates between saturation and cutoff. This signal is reversed by Tr2 and fed to the base of the main switching transistor Tr1 to drive it in the ON/OFF mode. In the ON state, current is drawn from the power supply to the load, while in the OFF state a circulating current occurs as shown by the chain line in the figure. Diode D1 is the free-wheeling diode for this current path. If the switching frequency is selected to be in the range of a few kHz to

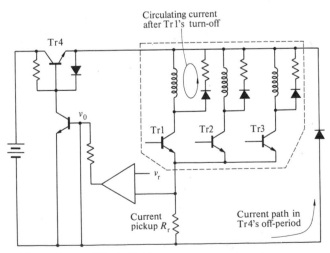

Fig. 5.33. A driver scheme of the PWM type.

30 kHz, the ripple component in the load current is very low. Since in this drive the voltage applied to the motor is sliced or chopped, this drive is often called the 'chopper'. When the current detected is smaller than the demanded value, the ON interval of Tr2 becomes longer than the OFF interval to draw more current from the supply. On the other hand, if the detected current is more than the demanded value corresponding to the reference voltage v_r, the OFF interval becomes longer than the ON interval to decrease the load current.

The mean voltage applied to the motor driver, E_L, is given by

$$E_L = E \times \text{time-ratio} = E \times \frac{t_{ON}}{t_{ON} + t_{OFF}}.$$

The advantages of the PWM or chopper drive are a single power supply, low power loss, and an automatically adjusted voltage to drive at a rated current.

Let us consider the current waveform in the circuit in Fig. 5.33. Just after Tr1 is turned on in the one-phase-on drive, the current is building up but is lower than the corresponding reference value including the superimposed AC component, and transistor Tr4 is in the ON state so that the high supply voltage is applied to the winding of Ph1 and the current builds up quickly. When the current enters the varying range around v_r/R_r, transistor Tr4 is driven in the ON–OFF or chopper mode, and the winding current will become such as shown in Fig. 5.34, and adjusted to the rated or demanded value. When Tr1 is turned off, the current decays quickly due to the diode/resistor suppressor put in parallel with the winding.

In the PWM driver of this type, the chopper frequency is determined by an external source. An alternative method is to incorporate a self-oscillation mechanism, making use of the electric time constant of windings. A disadvantage of the PWM is the electric and acoustic noise from the driver.

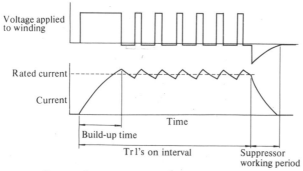

Fig. 5.34. Chopper voltage and current waveform.

5.4 Input controller

The last portion of the drive system to be discussed in this chapter is the input controller which governs the number of step command pulses and their timings, and in some application the direction signal.

5.4.1 Single-step controller

The simplest is the system which performs an increment with a single step. The step versus time relation in this system will be such as shown in Fig. 5.35. The positioning profile is generally oscillatory and its damping depends on the motor and drive scheme used. The input controller is very simple, since its function is only to provide an output signal which is suitable as the input signal to the sequencer. An example is shown in Fig. 5.36, which has the following features.

1. The input signal is clamped at a suitable H level. (In the figure at 5 V.)

2. Noise is absorbed in the condenser.

3. Since the input signal is deformed by the condenser, it is reformed by means of a Schmitt trigger. NAND or NOR gates may be used for the Schmitt trigger. If any part of the input signal can be less than the ground potential, a diode should be added as shown by the dotted curves. Most universal sequencers have a Schmitt trigger built in.

5.4.2 Input controller for electronic damper

To carry out a single-step without oscillation, a method called 'back-phasing'[1] is used. The relation between the position profile and pulse timing is illustrated in Fig. 5.37. The motor, at rest on an equilibrium position with Ph1 excited, is commanded to move to the next equilibrium position. If the rotor continues to be accelerated by the excitation of Ph2, it will overshoot exceeding the next equilibrium position. So, as the rotor is moving towards the next phase equilibrium position, Ph2 is switched off and Ph1 is switched back on. This produces a retarding torque which tends to slow down the rotor. When the rotor momentum is cancelled by

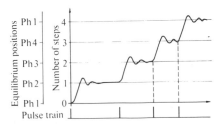

Fig. 5.35. Single-step response.

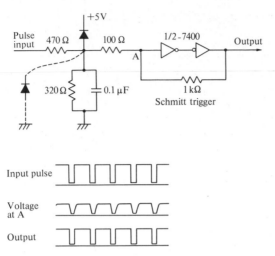

Fig. 5.36. Input controller used for the single-step operation, which may be widely used as a wave former.

the retarding torque, it will momentarily come to rest before reversing to go back to the previous position. At this moment, excitation is again switched to Ph2. The reversing pulse must be exactly timed so that the rotor reaches zero speed when it is on the equilibrium position of Ph2. Thus when the exciting current is switched back to Ph2, the rotor will lock in on the Ph2 equilibrium position with no overshoot or oscillations. A circuit of an input controller for back-phasing is shown in Fig. 5.38. This technique can be applied to more than one step of motion by the addition of more pulses preceding the back-phasing pulse. The adjustment of pulse positioning may be summarized as follows:

1. Adjust the pulse(s) preceding the back-phasing to give minimum response.

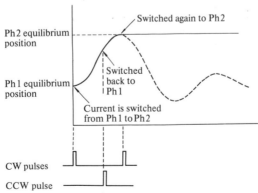

Fig. 5.37. Rotational angle and pulse timings in back-phasing damping.

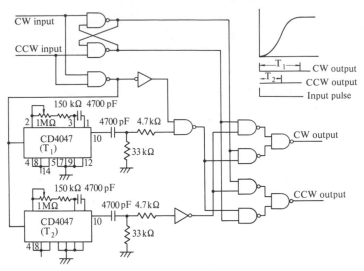

Fig. 5.38. An input controller for a back-phasing damped single-step motion.

2. Adjust the back-phasing pulse to retard the load motion such that it just reaches its final step position.

3. Adjust the last pulse to hold the load in its final position with minimum oscillation.

5.4.3 *Damped incremental motion with multi-steps*

Single-step motion is generally oscillatory. But non-oscillatory incremental motions can be performed with several steps by proper pulse timing. Two examples are given here.

(1) *Delayed-last-step electronic damping.*[1] This technique is explained by referring to Fig. 5.39. Assume it is wished to move three steps. If a three-pulse train is applied at a moderate rate the response will appear as shown in (a). However, if the period between the first and second pulses is adjusted such that the rotor will overshoot by exactly one step, its final step position will be Ph3. The last pulse is then applied to hold the rotor in place at its point of zero speed, as shown in (b). If the system friction is such that the rotor does not overshoot one step, this technique cannot be used. Likewise, this cannot be applied to less than three steps.

(2) *Constant-pulse-rate electronic damping.* If a non-oscillatory incremental motion is performed with several pulses at equal intervals, the input controller may be simple. Niimura[2],[4] presented examples of damped incremental motion on an outer-rotor motor of VR type, with

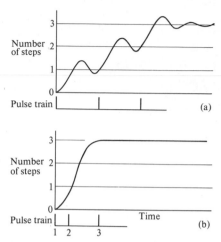

Fig. 5.39. (a) Ordinary three-step operation and (b) three-step DLSED.

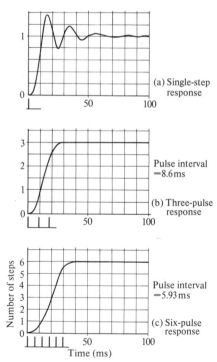

Fig. 5.40. Single-step motion and non-oscillatory motions with several pulses of equal interval.

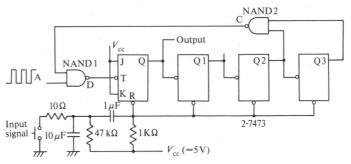

Fig. 5.41. An input controller to generate several pulses of equal interval.

three to six pulses of equal intervals in both single-phase-on and two-phase-on drive. Some examples of half-step-mode drive with 6 to 12 steps were presented too. Figure 5.40(a) shows its single-step response in the two-phase-on drive. Figures 5.40(b) and (c) are for a three-step and six-step drive, respectively.

Figure 5.41 illustrates a circuit to generate several pulses at uniform intervals. The waveforms in various parts of this circuit are illustrated in Fig. 5.42.

5.4.4 *Acceleration and deceleration*

To operate a stepping motor above its self-starting frequency, some acceleration/deceleration technique must be used to ensure that no loss of step occurs. In an open-loop drive system this means that the motor is started at a stepping rate at or below its maximum starting rate and then the stepping rate is increased with time until the desired speed is reached. Likewise, in most cases the motor must be decelerated to some speed below its maximum stop rate before it is stopped without positional error. Three typical techniques are presented here.

(1) *Gated oscillator.* An input controller of the gated oscillator type using a timing IC 555 is shown in Fig. 5.43, and the waveforms in the circuit are

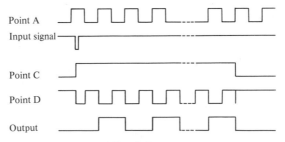

Fig. 5.42. Waveforms in the circuit of Fig. 5.41.

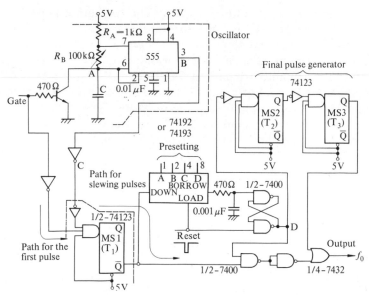

Fig. 5.43. A gated oscillator.

in Fig. 5.44. When the input signal reaches the L level, the first step-command pulse of width T_1 is injected by monostable chip MS1. On the other hand the potential at point A builds up after the input reaches the L level, and oscillations will be started in the 555 chip. The triangle waveform occurring at point A is shown in Fig. 5.44. The subsequent pulses to be used for slewing are injected at the minimum points in the

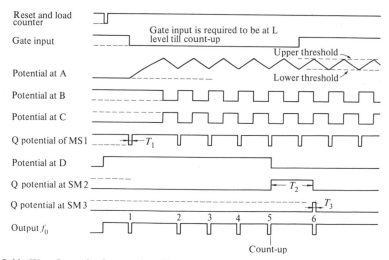

Fig. 5.44. Waveforms in the gated oscillator.

waveform, and the pulse width is also T_1. The oscillation period in this case is $0.7(R_A + R_B)C$. As is obvious from the figure, the oscillation period is shorter than the time interval between the first and second output pulses. If the number of pulses for one motion is N, $N-1$ should be preset in the counter, which is a 74192 or 74193 type here. When counted up, the potential at point D reaches the L level to start mono-stable chip MS2, and at time T_2 a pulse of width T_3 is injected by MS3. The time T_2 is set longer than the oscillating period in slewing for the sake of successful stopping. In some cases of frictional load, however, it is not necessary to set the last step interval longer than before. The MS2 and 3 stage is not needed in this case.

(2) *RC acceleration.* When the slewing rate is much higher than the pull-in rate, a more sophisticated method is required for acceleration of a stepping motor. If the motor/load combination produces torque decreasing with speed, the ideal acceleration ramp is one in which the rate of acceleration decreases with stepping rate, i.e. speed is increased rapidly at low stepping rates, and increased less rapidly at high stepping rates. A method of accomplishing this by means of conventional techniques is shown in Fig. 5.45. When a start signal is applied to the input terminal, an exponential potential curve appears across the condenser C, the time constant being $C(R_1 + R_2)$. This is used as a part of the input signal to the V–F converter to generate acceleration pulses. The input signal to the V–F converter for deceleration is produced by discharging the condenser through the resistor R_3, which is started with a deceleration-start signal

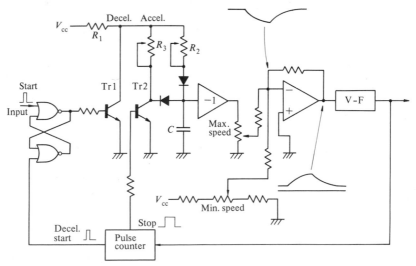

Fig. 5.45. Block diagram for the RC ramping.

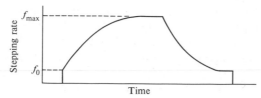

Fig. 5.46. Pulse frequency profile in the RC ramping.

injected by the pulse counter. When the required number of pulses have been generated by the V–F converter, the pulse counter sends a signal to the base of Tr2 to discharge the condenser at once. The maximum and minimum speeds are adjustable by the two variable resistors R_2 and R_3. Figure 5.46 illustrates the curve of pulse frequency vs. time in RC acceleration/deceleration.

One of the disadvantages with this technique is that the rate of deceleration can be too high at the very beginning of deceleration. An improvement is discussed in Reference [1].

5.5   Throw-up and throw-down control by a microprocessor

In this section a linear acceleration/deceleration control using a microprocessor will be discussed. One of the most remarkable features of the microprocessor lies in its flexibility in software. A microprocessor also has an advantage in that it can perform not only the generation of pulse timings but also logic sequencing and the role of the input controller. The program here presented is for a simple throw-up and throw-down control. But a more sophisticated speed profile can be realized by proper programming.

5.5.1   *Basic concept of using an 8080-family microprocessor and general flow-chart*

There is a great variety of designs of suitable software for a particular type of microprocessor. In the present example, a simple algorithm using an 8080A microprocessor will be presented. This can be applied to any 8080-family processors. The CPU, or central processing unit, of an 8080A has six registers which can be used for general purposes. We use them for the following roles, respectively.

Register B. This register is used to store the present excitation state of the windings.

Registers C and E. Register C is used here for counting the completed steps, and register E for storing the number of remaining steps in a motion. When a motion instruction of p steps is given, the number p is loaded in register E, and C is cleared to zero. After a step-command signal is sent out, C is incremented and E is decremented by one. An example of the relation between pulse intervals and the changes of the contents of registers C and E is shown in Fig. 5.47 for several cases of steps. The commanded speed versus time profiles may be illustrated as in Fig. 5.48; the motor is instructed to start at stepping rate f_0, accelerate linearly up to the slewing rate f_s, and run at this constant rate. Before stopping, the motor is decelerated and then instructed to stop from a speed at which the motor can stop without an overrun.

Calculations of pulse intervals and numbers of pulses used before reaching the slew rate are discussed in Section 6.4. In the program presented here the motor is accelerated with six pulses, and the deceleration is performed using the same pulse intervals used for acceleration, in the reverse order. In the examples in Fig. 5.47, the motions with less than 13 steps do not have a slewing region; the motor is accelerated and then decelerated. Motions with 13 or more steps have a slewing region in which step-command pulses are given at a constant rate.

As will be explained later, the contents in registers C and E are used for computing the addresses of memories storing the data from which proper pulse timings are produced.

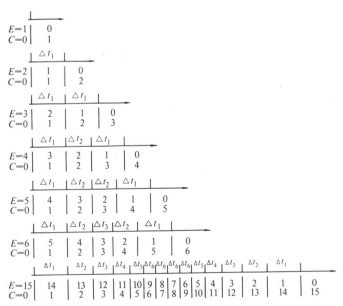

Fig. 5.47. Relation between pulse intervals and changes of data in registers E and C.

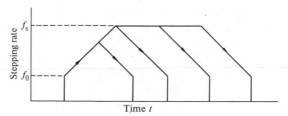

Fig. 5.48. Commanded speed vs. time profile relation.

Register D. This is used to store the code which governs the direction of rotation as follows:

$D=0$ or 00000000 for CW direction

$D=1$ or 00000001 for CCW direction.

We will arrange it so that if $D=2$, the processor judges that this is the end of the data for rotation; thus

$D=2$ or 00000010 for END of program.

The instructions on direction and number of steps to rotate in a motion are stored in a memory area starting from address Y as shown in Fig. 5.49. Before the first motion the datum in address Y is transferred to register D, and that in $Y+1$ to register E. In the second motion the data in addresses $Y+2$ and $Y+3$ are transferred to registers D and E, respectively.

Address	Memory byte	
Y		for the 1st motion
Y+1		
Y+2		2nd
Y+3		
Y+4		3rd
Y+5		
Y+6		4th
Y+7		
Y+8		5th
Y+9		
Y+10		6th
Y+11		
Y+12		7th
Y+13		
Y+14		
Y+15		
Y+16		

Fig. 5.49. Memory area for direction and number of steps to rotate.

Address   Memory byte

Address	Memory byte	
X−1		
X	Q_1	for Δt_1
X+1	Q_2	for Δt_2
X+2	Q_3	for Δt_3
X+3	Q_4	for Δt_4
X+4	Q_5	for Δt_5
X+5	Q_6	for Δt_6
X+6		
X+7		

Fig. 5.50. Memory area for pulse interval data.

Pair register HL. In the software for an 8080 microprocessor, the pair register HL is often used to specify the memory address from (or to) which a datum is transferred to (or from) a register. In the present program the HL register is used to specify the memory addresses storing the pulse interval data. Before loading the HL register, a computation to determine the correct address must be done. Figure 5.50 illustrates the memory area for the pulse-interval data, which is composed of six addresses from X to X+5. (The address X is specified by an assembler program as a suitable number.) It is naturally possible to expand this area when more steps are used in acceleration and deceleration. The real pulse interval Δt_m is computed from the datum Q_m in each memory byte and the number of states in execution of some part of the program, and is given by

$$\Delta t_m = a Q_m + b$$

where a and b are constants which are determined by the software and the clock frequency in the microprocessor.

Memory addresses are specified downwards from X in acceleration, and upwards returning to X in deceleration. The slewing-pulse interval is generated using the datum Q_6 stored in address X+5.

The pair register HL is also used for specifying the addresses of data on direction and number of steps to rotate.

5.5.2 *Flow-chart and program*

The overall flow-chart of the program to be discussed is illustrated in Fig. 5.51. The details of each part are given in other figures incorporating assembler-language codings. The explanation of the flow chart and program will be given part by part as follows:

(1) *START to PUSH HL.* The detail of this part is illustrated in Fig. 5.52. The step-by-step explanation is as follows:

START   Program starts from address 8200H in this example. The symbol H indicates that the preceding number is a hexadecimal number.

B ← 00110011   Register B is initialized and the excitation state is reset. This stage is composed of three stages:

(i) A ← 00110011. Accumulator AC is loaded with 00110011, which is 33 in hexadecimal expression, to drive a motor in the two-phase-on mode.

(ii) SEND A. This code is sent out to the driver through an output port as shown in Fig. 5.53. The windings of Ph1 and Ph2 are excited, since a positive

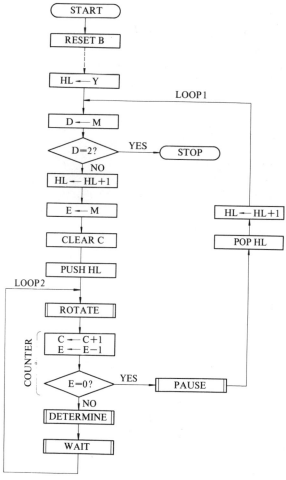

Fig. 5.51. Overall flow-chart of the program.

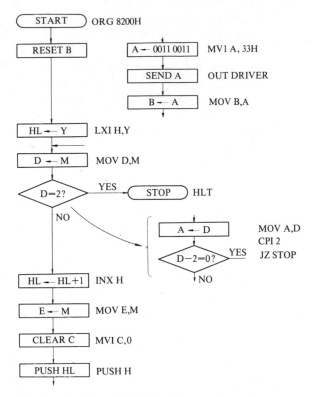

Fig. 5.52. Detailed flow-chart from START to PUSH HL.

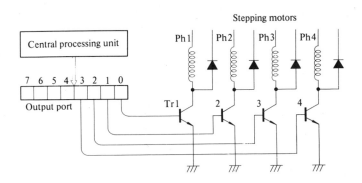

Fig. 5.53. Connection between the output port and driver. (See Fig. 5.14 for details.)

potential of 5 V is applied to the base circuits for Ph1 and Ph2.

(iii) B ← A. The datum in AC is transferred in register B to be stored in the CPU.

HL ← Y   Pair register HL is loaded with number Y which is the starting address of the memory area storing the data on direction and number of steps. As shown in Fig. 5.49 one motion is instructed by two bytes; the first byte is for the direction code and the second byte for the number of steps.

D ← M   The datum in memory address Y is transferred to register D.

D = 2?   If the datum is 2 the program is branched off to END which is instructed by the mnemonic code HLT.

HL ← HL + 1   If D = 2, the datum in HL is incremented by one in the next stage.

E ← M   The datum in address Y + 1 is transferred to register E.

C ← 0   The content of register C is cleared to zero.

PUSH HL   The datum in HL is pushed down, since the HL register will be used for another purpose.

(2) *Logic sequencer* (*ROTATE routine*). Before discussing the generation of switching signals in the programmed timings, we shall see, in Fig. 5.54, how the logic sequencing mechanism can be carried out inside the central processing unit (CPU).

D = 0?   The datum in register D is transferred to AC (= accumulator) and it is compared with zero (= 00H). If D equals zero the program proceeds to the RIGHT routine, and if D is not zero, that is 1, the program proceeds to the LEFT routine.

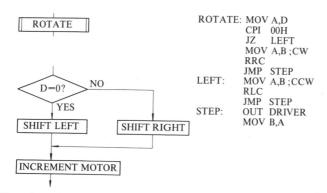

```
ROTATE: MOV A,D
        CPI  00H
        JZ   LEFT
        MOV  A,B ;CW
        RRC
        JMP  STEP
LEFT:   MOV  A,B ;CCW
        RLC
        JMP  STEP
STEP:   OUT  DRIVER
        MOV  B,A
```

Fig. 5.54. Flow-chart of ROTATE routine.

SHIFT RIGHT In the RIGHT routine, the datum in register B is transferred to AC and it is rotated right by one bit by the mnemonic instruction RRC; the datum in AC is changed to 10011001.

SHIFT LEFT In the LEFT routine, the eight-bit datum in AC is rotated left by one bit by the mnemonic instruction RLC.

INCREMENT MOTOR The datum in AC is sent out through the output port to advance the motor by one step angle clockwise or counter-clockwise. Finally the datum in AC is stored in register B.

The change of the datum in register B is illustrated in Fig. 5.55(a). It is seen that the logic sequencer function is implemented by taking the lower four bits out of the eight-bit signal on the output port. Figure 5.55(b) is for the single-phase excitation; the lower four bits are taken, too, but the initial data for register B is 00010001. Figure 5.55(c) is for the half-step-drive case. The initial datum for B is 00000111 and the four even-numbered bits are used for excitation signals.

(3) *COUNTER routine.* The details of step counting and Loop 1 are illustrated in Fig. 5.56. The counter function is as follows:

$C \leftarrow C + 1$ As register C is used as the step counter, the datum in
$E \leftarrow E - 1$ this register is incremented after a step. Register E is decremented.

$E = 0?$ If register E becomes zero, which means that a motion is completed, the program is branched off to LOOP1. Otherwise the program will proceed to the next stage.

(4) *PAUSE routine.* In LOOP1 the PAUSE routine is first executed; this is the routine which waits for an interval of time for a system to perform a

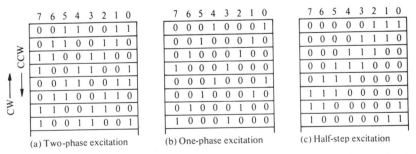

Fig. 5.55. Change of datum in register B; (a) two-phase excitation, (b) single-phase excitation, and (c) half-step excitation.

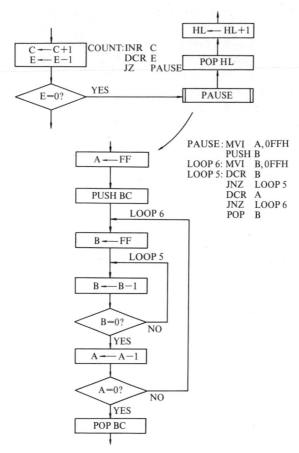

Fig. 5.56. COUNTER routine.

function such as to print a character on a printer. The details are as follows:

A ← FF   AC is loaded with FF. The hexadecimal FF equals binary 11111111 or decimal 255.

PUSH BC   Register B was used before for storing the excitation state information. In order to use this register for another purpose here, the content of this register is pushed or stored in the memory address specified by the stack pointer. In a 8080 processor, PUSH BC pushes the contents of the register pair B and C.

B ← FF   Register B is loaded with FF.

B ← B − 1 ⎫
B = 0?   ⎬   The content of B is repeatedly decremented until it becomes 0.
              ⎭

$A \leftarrow A - 1$ }
$A = 0$? }   After decrementing B to 0, the content of AC is repeatedly decremented until it becomes 0. This process takes a comparatively long time, since loading FF on register B and its decrementing to 0 takes place each time.

POP BC   Since an interval of time is consumed in the above process, register B recovers the previous memory on excitation state. Register C recovers its previous content, too.

This routine goes through 995 868 states. Hence, if the time spent in one state is 0.5 μs, the total time used in the routine is about 0.5 s. By loading register B with a smaller number, the total wait time will be shorter.

(5) *Determine memory address for pulse interval datum.* If the content of register E is not zero, the program proceeds to the next stage or DETERMINE routine. In this routine, the memory address storing the

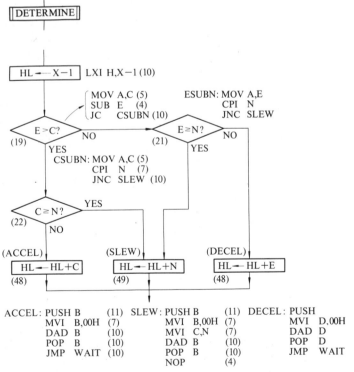

Fig. 5.57. DETERMINE routine.

datum to produce the proper wait interval before the next switching signal is computed from the data in registers C and E, and it is stored in the pair register HL. The details of this routine are shown in Fig. 5.57. First HL is initialized by being loaded with $X-1$; here X is the first address of the memory area containing data for producing pulse intervals (see Fig. 5.50). The memory address can be specified by adding a number, selected from 1 to 6, to the initial value $X-1$ in the pair register HL. The computation of a number to be added is carried out on the following principles.

(i) First it is decided whether the motion is more than half complete. If $E>C$, the motion is in the first half, and if $C\geq E$, the second half.

(ii) In the first half. It is decided whether C is larger than or equal to 6, or C is smaller than 6. It is seen in Fig. 5.47 that if $C<6$, the motion is on acceleration and the subscript m for Δt_m is equal to C. Consequently the number to be added to $X-1$ is C.

(iii) In the second half. It is decided whether E is larger than or equal to 6, or E is smaller than 6. It is seen in Fig. 5.47 that if $E<6$, the motion is decelerating and the subscript m for Δt_m is equal to E. Consequently the number to be added to $X-1$ is E.

(iv) If $C\geq 6$ in the first half or $E\geq 6$ in the second half, the motion is slewing. Consequently the number to be added to $X-1$ is 6.

In the assembler-language coding the number 6 is denoted by N. It is possible to equate N to another number when more or less pulses should be used before entering slewing.

The numbers in parentheses in Fig. 5.57 are those of states in processing, which are important factors to be taken into account for calculating the data which are stored in memory addresses for time intervals.

(6) *WAIT routine.* The next stage is the routine to produce the pulse interval or the interval between the previous and next switching signals. As shown in Fig. 5.50, the memory area from X to $X+5$ is used for storing the pulse interval data.

The flow-chart of Fig. 5.58 illustrates the details of this routine as explained in the following.

$A \leftarrow Q$	The datum stored in the memory address specified by HL is transferred to AC.
$A \leftarrow A-1$	The datum in AC is decremented repeatedly until it
$A = 0?$	becomes zero. A certain time is spent in this process. When AC becomes zero the execution of this routine ends, and the program will return to the ROTATE routine to send out the next switching signal to rotate the motor through another step angle.

5.5.3   *Calculation of numbers for producing pulse intervals*

The next problem is how to determine the data to be stored in the memory area starting from X. This is determined in connection with the number of states in each routine and the time spent in each state. The numbers in parentheses under blocks in Fig. 5.58 are numbers of states in an 8080 microprocessor. The total number of states in the WAIT routine is $15Q+2$, where Q is the datum transferred from a memory address to AC. The number of states in the ROTATE routine is 56 and that in the counter stage is 20. Finally the number of states in the DETERMINE routine is 99(100); the number in parentheses is that for slewing. Hence the total number S of states required between two subsequent switching signals is

$$S_m = 15Q_m + 177 \,(178).$$

We may approximate 178 to 177, for simplicity in the subsequent calculations.

If one state spends 0.5 μs, the pulse interval is given by

$$\Delta t_m = (15Q_m + 177) \times 0.5 \times 10^{-6} \,\text{(s)}.$$

Table 5.2 shows an example of pulse intervals and the numbers Q_m from which the pulse intervals are produced in the present program.

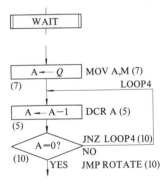

Fig. 5.58. WAIT routine.

Table 5.2. Examples of pulse intervals and data to produce pulse intervals.

m	Pulse interval Δt_m (ms)	Q_m	Integer approximation	Hexadecimal expression
1	1.984	252.2	252	FC
2	1.460	182.9	183	B7
3	1.212	149.8	150	96
4	1.059	129.4	129	81
5	0.952	115.1	115	73
6	0.873	104.6	105	69

Table 5.3. Overall program in 8080 assembler language.

Label	Code	Operand	
	ORG	8200H	;Start from address 8200
RESET:	MVI	A,33H	;Accumulator is loaded with 00110011
	OUT	DRIVER	;Datum in Accumulator is sent out
	MOV	B,A	;Datum in Accumulator is transferred to register B
	LXI	H,Y	;Pair register HL is loaded with address Y
LOOP1:	MOV	D,M	;Datum in address Y is transferred to register D
	MOV	A,D	;Datum in register D is transferred to Accumulator
	CPI	2	;A − 2
	JZ	STOP	;If A = 2, jump to address STOP
	INX	H	;HL ← HL + 1
	MOV	E,M	;Datum in address Y + 1 is transferred to register E
	MVI	C,0	;Register C is cleared
	PUSH	H	;Data in HL are pushed down
ROTATE:	MOV	A,D	;Datum in register D is transferred to Accumulator
	CPI	0	;A − 0
	JZ	LEFT	;If A = 0, jump to address LEFT
RIGHT:	MOV	A,B	;Datum in register B is transferred to Accumulator
	RRC		;A is rotated right
	JMP	STEP	;Jump to address STEP
LEFT:	MOV	A,B	;Accumulator is loaded with B
	RLC		;A is rotated left
	JMP	STEP	;Jump to address STEP
STEP:	OUT	DRIVER	;A is sent out
	MOV	B,A	;Register B is loaded with A
COUNT:	INR	C	;Register C is incremented
	DCR	E	;Register E is decremented
	JZ	PAUSE	;If E = 0, jump to address PAUSE
DETERM:	LXI	H,X−1	;Register pair HL are loaded with X − 1
	MOV	A,C	;
	SUB	E	;C − E
	JC	CSUBN	;If E > C, jump to CSUBN
ESUBN:	MOV	A,E	;
	CPI	N	;E − N
	JNC	SLEW	;If E ≥ N, jump to SLEW
DECEL:	PUSH	D	;⎫
	MVI	D,0	;⎬ HL ← HL + E
	DAD	D	;⎪
	POP	D	;⎭
	JMP	WAIT	;Jump to WAIT
CSUBN:	MOV	A,C	;
	CPI	N	;C − N
	JNC	SLEW	;If C ≥ N, jump to SLEW
ACCEL:	PUSH	B	;⎫
	MVI	B,0	;⎬ ← HL + C
	DAD	B	;⎪
	POP	B	;⎭
	JMP	WAIT	;
SLEW:	PUSH	B	;⎫
	MVI	B,0	;⎪
	MVI	C,N	;⎬ ← HL + N
	DAD	B	;⎪
	POP	B	;⎭
	NOP		;No operation for adjustment of number of states
WAIT:	MOV	A,M	;WAIT routine

Table 5.3. *Contd.*

Label	Code	Operand	
LOOP4:	DCR	A	;A ← A − 1
	JNZ	LOOP4	;If A ≠ 0, jump to LOOP4
	JMP	ROTATE	;Jump to ROTATE routine
PAUSE:	MVI	A,0FFH	;PAUSE routine, A ← 11111111
	PUSH	B	
LOOP6:	MVI	B,0FFH	;B ← 11111111
LOOP5:	DCR	B	;B ← B − 1
	JNZ	LOOP5	;If B ≠ 0, jump to LOOP5
	DCR	A	;A ← A − 1
	JNZ	LOOP6	;If A ≠ 0, jump to LOOP6
	POP	B	;End of PAUSE routine
	POP	H	;Recover memory address for motion instruction data
	INX	H	;HL ← HL + 1
	JMP	LOOP1	;Jump to LOOP1
STOP:	HLT		
DRIVER	EQU	2	;Output port 2 is used for output for motor driver
N	EQU	6	;N is set to 6
X:	DB	252	;Datum to generate Δt_1
	DB	183	;Datum to generate Δt_2
	DB	149	;Datum to generate Δt_3
	DB	129	;Datum to generate Δt_4
	DB	115	;Datum to generate Δt_5
	DB	104	;Datum to generate Δt_6
Y:	DB	1	;⎫ Direction and number of steps for the
	DB	10	;⎭ first motion (CCW 10 steps)
	DB	1	;⎫ Direction and number of steps for the
	DB	5	;⎭ 2nd motion (CCW 5 steps)
	DB	0	;⎫ Direction and number of steps for the
	DB	23	;⎭ 3rd motion (CW 23 steps)
	DB	0	;⎫ Direction and number of steps for the
	DB	15	;⎭ 4th motion (CW 15 steps)
	DB	1	;⎫ Direction and number of steps for the
	DB	3	;⎭ 5th motion (CCW 3 steps)
	DB	0	;⎫ Direction and number of steps for the
	DB	33	;⎭ 6th motion (CW 33 steps)
	DB	1	;⎫ Direction and number of steps for the
	DB	18	;⎭ 7th motion (CCW 18 steps)
	DB	0	;⎫ Direction and number of steps for the
	DB	5	;⎭ 8th motion (CW 5 steps)
	DB	1	;⎫ Direction and number of steps for the
	DB	11	;⎭ 9th motion (CCW 11 steps)
	DB	0	;⎫ Direction and number of steps for the
	DB	60	;⎭ 10th motion (CW 60 steps)
	DB	1	;⎫ Direction and number of steps for the
	DB	29	;⎭ 11th motion (CCW 29 steps)
	DB	1	;⎫ Direction and number of steps for the
	DB	9	;⎭ 12th motion (CCW 9 steps)
	DB	0	;⎫ Direction and number of steps for the
	DB	1	;⎭ 13th motion (CW 1 step)
	DB	1	;⎫ Direction and number of steps for the
	DB	25	;⎭ 14th motion (CCW 25 steps)
	DB	0	;⎫ Direction and number of steps for the
	DB	9	;⎭ 15th motion (CW 9 steps)
	DB	2	; End of data to be printed
	END		

5.5.4   *Overall program*

The overall program expressed in the 8080 assembler language is given in Table 5.3. The grammar of the assembler language is given in Reference [5]. This program can be translated into the machine language either manually or by a cross-assembler using a computer. The meaning of each step is stated after a semicolon (;).

In this example, fifteen motions are instructed using thirty memory bytes starting from Y.

References for Chapter 5

[1] Maginot, J. and Oliver, W. (1974). Step motor drive circuitry and open loop control. *Proc. Third Annual Symposium on Incremental motion control system and devices.* Department of Electrical Engineering, University of Illinois, pp. B1–39.
[2] Kenjo, T. and Niimura, Y. (1979). *Fundamentals and applications of stepping motors.* (In Japanese.) p. 157. Sogo Electronics Publishing Co., Ltd., Tokyo.
[3] Pawletko, J. P. and Chai, H. D. (1976). Three-phase variable-reluctance step motor with bifilar winding. *Proc. Fifth Annual Symposium on Incremental motion control system and devices.* Department of Electrical Engineering, University of Illinois, pp. F1–8.
[4] Niimura, Y. (1974). Outer-rotor-type stepping motor. *Proc. Third Annual Symposium on Incremental motion control system and devices.* Department of Electrical Engineering, University of Illinois, pp. H1–10.
[5] Intel 8080 Assembly Language Programming Manual. Intel Corporation, Santa Clara, California (1975).

6. Torque characteristics and pulse intervals—measurement and use in system design

In this chapter, methods of measuring the torque characteristic curves of a stepping motor will first be given, then follows a discussion on the relation between the torque curves, dynamic equation, and acceleration. Lastly, several theories for determining pulse intervals for acceleration and deceleration are developed for use in system design.

6.1 Measurment of static characteristics

6.1.1 T/θ characteristics and holding torque

The test motor is kept stationary by supplying rated current in a specified excitation scheme, say, one-phase or two-phase excitation. One of the three or four phases shall be arbitrarily chosen to be supplied the current in the case of one-phase-on scheme. The rotor position at no load is defined as the 'equilibrium' or 'rest position'. We are now going to measure the relationship between the torque applied to the shaft and displacement from the equilibrium position. As illustrated in Fig. 6.1 external torques are provided by means of a pulley, a string, and a weight. When the weight is zero, the rotor is at rest at the equilibrium position, and as the weight is increased gradually in the order of W_1, W_2, $W_3 \cdots$, the rotor moves clockwise following the positions θ_1, θ_2, $\theta_3 \cdots$. Figure

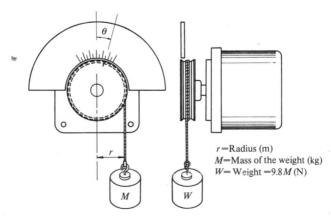

r = Radius (m)
M = Mass of the weight (kg)
W = Weight = $9.8M$ (N)

Fig. 6.1. Arrangement to measure T/θ characteristics.

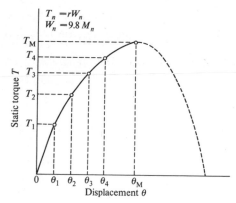

Fig. 6.2. A T/θ characteristic curve.

6.2 is a graph plotting torques $T_i = rW_i$ versus displacements θ_i. As displacement is increased, the torque eventually reaches a maximum at displacement θ_M. The T/θ characteristics for displacements larger than θ_M, which are shown by a dotted curve in Fig. 6.2, cannot be successfully measured by the apparatus shown in Fig. 6.1.

Ideally the T/θ characteristic curves for each phase (in the one-phase excitation) or each combination of phases (in the two-phase excitation) are identical. But in most practical motors, variations are observed as illustrated in Fig. 6.3. The holding torque must be defined as the minimum of the maximum static torques. This definition of the holding torque is effectively identical to the definition given in Section 2.5.1: the maximum static torque that can be applied to the shaft of an excited motor without causing continuous motion. In Fig. 6.3 the rotor position is taken with respect to an equilibrium position of Ph1, and it is seen that

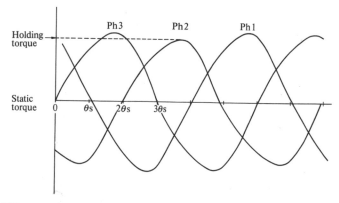

Fig. 6.3. T/θ curves for three phases.

the equilibrium positions for the other phases are not on the rated positions $2\theta_s$, $4\theta_s$, etc. The definition of positioning accuracies was given in Section 2.1.3.

An optical encoder or a precision contactless potentiometer can also be used as another method for measuring displacements.

6.1.2 *T/I characteristics*

Generally, the maximum static torque increases with current. The plots of maximum static torque versus current in the specified excitation scheme are called the *T/I* characteristics, and are as shown in Fig. 6.4. This figure compares the cases of a VR motor and a hybrid motor, both being of 1.8° step angle. The torque in a VR motor is zero when it is not excited, and increases with current parabolically in low current ranges. The rate of increase in the higher current ranges is small due to magnetic saturation in the cores. In a hybrid motor or a PM motor, a static torque appears even if it is not excited, and this specific torque is referred to as 'detent torque'. The manner of increase of static torque with current is close to a linear relation in a hybrid motor and a PM motor.

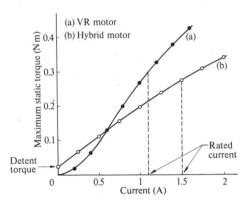

Fig. 6.4. *T/I* characteristic curve.

6.2 Measurement of dynamic characteristics

The dynamic characteristics of a stepping motor involve the pull-in characteristics and pull-out characteristics.

6.2.1 *Importance of couplings in dynamic characteristics*

Measurement of dynamic characteristics is affected strongly by electronic and mechanical conditions in the apparatus, since, as discussed in Chapter 4, the dynamic behaviour of a stepping motor varies with load inertia and the damping mechanism incorporated both in the mechanical and elec-

Rubber coupling   Bellows coupling   Direct coupling

Fig. 6.5. Couplings.

tronic set-up in a drive system. When two or more parties, say a manufacturer and a user, discuss details of the dynamic characteristics of a stepping motor, it is first of all necessary to specify the measuring method and drive circuit.

One of the most important mechanical factors in the measurement apparatus is the coupling between the motor shaft and torque meter. Figure 6.5 illustrates three typical couplings: rubber coupling, bellows coupling, and direct coupling. Figure 6.6 illustrates a variation appearing in the pull-out torque curves with differences in couplings used, measured by Niimura.[1] The test motor was a four-phase hybrid motor and it was

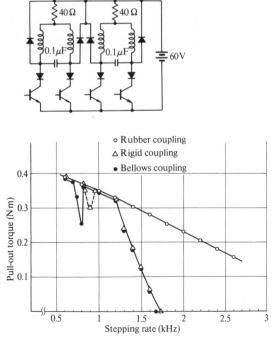

Fig. 6.6. Variation of pull-out torque curves with differences of couplings. (After Ref. [1].)

driven in the two-phase-on scheme using the drive shown in the same figure. The torque meter used is a hysteresis dynamometer (shown in Fig. 6.18). Excellent pull-out characteristics are obtained with a rubber coupling in a wide range of stepping rates, since rotor oscillations are absorbed in the flexible rubber used in the coupling. On the other hand, when a rigid or bellows coupling is used, the pull-out torque decreases suddenly in the range past 1200 Hz, and becomes zero at about 1700 Hz. These poor characteristics are presumably due to the coupling mechanism which is not capable of absorbing rotor oscillations.

Backlash in the coupling is another mechanism which absorbs oscillations and also an important factor which affects the pull-out curve. The effects of backlash on the single-step response and pull-out characteristics are reported by Ward and Lawrenson,[2] and it is shown that in the single-step response an improved response is obtained by the use of backlash both in reducing overshoot and in increasing the rate of decay of oscillation. For the comparison of rigid couplings and couplings with backlash in the pull-out characteristics they presented Fig. 6.7, which was obtained using a VR motor. With a rigid coupling a dip appears around a stepping rate of 140 Hz due to oscillations at the natural frequency, and the high speed response terminates above 1300 Hz. With backlash more dips occur at lower stepping rates, and, around 160 and 220 Hz, the test motor is not capable of stable running until significant load torque is applied. It should be noted, however, that the maximum pull-out stepping rate is increased due to backlash between the motor and load.

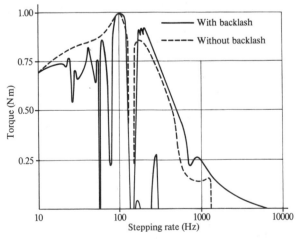

Fig. 6.7. Pull-out torque versus speed curves comparing the two cases: with backlash and without backlash.

6.2. *Pull-in torque characteristics*

The pull-in torque characteristics or starting characteristics are the plots of range of frictional load torque at which the motor can start and stop without losing synchronism for various frequencies. The pulse train used in the measurement is such as shown in Fig. 6.8 and the number of pulses involved is usually 100 or 200. When measuring the pull-in range, it is very important to specify the inertia coupled to the motor shaft as well as the coupling type and driver circuit. The pull-in range decreases with load inertia as illustrated in Fig. 6.9. In the pull-in characteristics, not only the maximum load torques but also the minimum load torques must be specified. For example, in Fig. 6.10, at a pulse rate around f_0, the motor is not capable of starting due to oscillation until a significant frictional torque is applied.

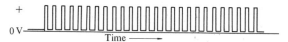

Fig. 6.8. A pulse train of constant frequency.

It should be noted that the pull-in torque curves vary with how the braking torque is applied. In the case of Fig. 6.11, the gravitational force applied to the load object always gives an external torque to the motor shaft. Let us assume that Ph1 is excited in the beginning and an external torque T_1 is applied to the shaft causing a displacement θ_1 from the equilibrium position (see Fig. 6.12). As soon as the first pulse is applied to the driver, Ph1 is turned off and Ph2 is, instead, turned on, and the motor torque becomes T_1' which is larger than T_1. The difference $T_1' - T_1$ is used to accelerate the motor. If this torque difference is large enough to accelerate the load and synchronize with the following pulses, T_1 is in the pull-in range. As T_1 increases causing displacement θ_1 to be larger, the

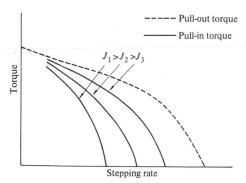

Fig. 6.9. Variation of pull-in ranges with load inertia.

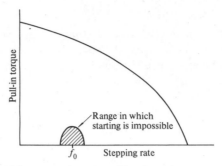

Fig. 6.10. Range in which a motor is not capable of starting.

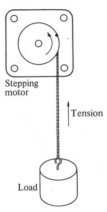

Fig. 6.11. Constant braking torque due to gravity.

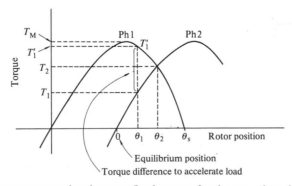

Fig. 6.12. Motor torque causing the motor/load to start for the case of gravitational load.

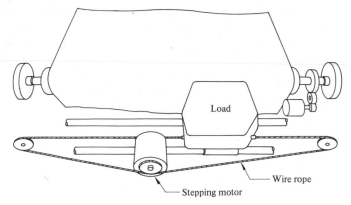

Fig. 6.13. A system in which the load torque at starting is a static frictional torque.

torque difference becomes smaller and the maximum pulse rate at which the motor can start becomes lower. At $\theta = \theta_2$ the two curves intersect. The torque produced at this point is the maximum pull-in torque, since the torque difference is zero and the motor can start only at an extremely low pulse rate. It is concluded from Fig. 6.12 that there is a significant difference in the maximum pull-in torque and maximum static torque or holding torque. In the sine-curve model of the static torque plots against rotor position, the ratio of the holding torque to the maximum pull-in torque is $\sqrt{2}$.

In the case of the load shown in Fig. 6.13 the load torque appearing at starting is the static frictional torque. This means that the initial position of the rotor can be any position in a range centring on the equilibrium position, in which the absolute value of the motor's static torque is less than the static frictional torque of the load (see Fig. 6.14). In this figure,

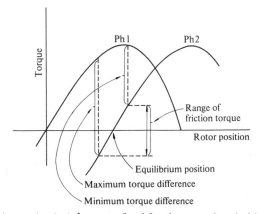

Fig. 6.14. Motor torque to start the motor/load for the case of static frictional load.

the farther the initial position is towards the left from the equilibrium position, the more the torque difference which appears on switching excitation, and the higher the pull-in capability. On the other hand, the farther towards the right the initial position is from the equilibrium position, the less the torque difference. In giving the standard specifications for pull-in torque, measurements should be made in the worst conditions that produce the smallest torque differences.

6.2.3 *Pull-out torque curves*

The test motor is started in the pull-in range and accelerated at a light load up to the stepping rate at which the pull-out torque is to be measured. When the load torque is increased gradually at this speed, the motor will eventually loose synchronism. This load torque is the pull-out torque at the stepping rate. If the pull-out torques are plotted against stepping rate, a graph like Fig. 6.6 or 6.7 will be obtained. Generally, the pull-out torques at very low stepping rates are close to the pull-in torques, as shown in Fig. 6.15.

There are several means of applying load torques in measuring pull-out torques. The most basic principle is the prony (or cord and pulley) brake and the load torques are measured by one or two spring scales as illustrated in Fig. 6.16(a) and (b). The load torques T_L are determined by the following equations respectively.

(a) Two-scale method

$$T_L = R(q_1 - q_2) \quad \text{(N m)} \tag{6.1}$$

where R = pulley radius (m)
q_1 = reading on scale 1 (N)
q_2 = reading on scale 2 (N).

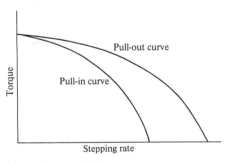

Fig. 6.15. Pull-in and pull-out torque curves.

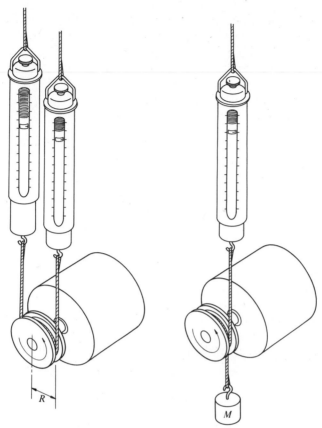

Fig. 6.16. Measuring motor torques by the use of spring scales; (a) two-scale method and (b) one-scale method.

(b) One-scale method

$$T_L = R(9.8M - q) \ (\mathrm{N\,m}) \tag{6.2}$$

$M = $ mass of the weight (kg)

$q = $ reading on the scale (N).

Strain gauges are frequently used for measuring torques, for example in the apparatus illustrated in Fig. 6.17.[3]

The thumb-screw sets the tension on the cord which controls the friction torque, and strain gauges are connected in a bridge circuit to read the difference in tensions $T_1 - T_2$. This apparatus may be used for measurement of the pull-in torques as well. The initial load situation is, however, similar to that in Fig. 6.13.

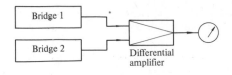

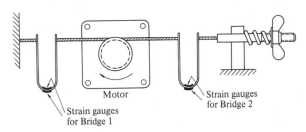

Fig. 6.17. Measuring motor torques by the use of strain gauges.

The hysteresis dynamometer (Fig. 6.18) which is an instrument de-signed for dynamic torque measurements, may be used for measurement of pull-out torques with a stepping motor. But this torque meter is not always suitable for a stepping motor having a small step angle due to the cogging in the torque production mechanism employing the hysteresis phenomenon.[3]

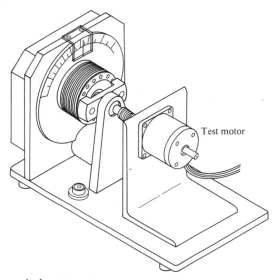

Fig. 6.18. A hysteresis dynamometer.

6.3 Dynamic equation and acceleration

6.3.1 *Dynamic equation*

When a stepping motor is synchronizing with a pulse train, the torque produced by the motor is equal and opposite to the load torque which is the sum of the torque to accelerate the rotor/load inertia and the frictional torque. This statement may be expressed by the fundamental dynamic equation:

$$\tau_M = J\frac{d\omega}{dt} + D\omega + T_f \qquad (6.3)$$

where τ_M = torque produced by the rotor
J = inertia of rotor and load combination
ω = angular speed of rotor
D = viscous frictional constant
T_f = frictional load torque independent of speed.

The motor torque τ_M is a function of speed, magnetomotive force, torque angle and other machine parameters as in the stationary torque discussed in Section 4.4.1. However it is here treated as the torque which is produced by a specific motor driven by a specific driver circuit in a specific excitation mode. In using eqn (6.3) we assume that: (i) a mechanical damper is not used; (ii) the motor torque does not have any oscillating component in the speed ranges under consideration.

The first term on the right-hand side is the torque required to accelerate the inertia of rotor and load combination. When the rotor torque is transmitted to the load by means of gears, belts, or the like, the inertia J is not the inertia of load itself, but must be the quantity reflected at the shaft. Some formulae for reflected inertia are shown in Appendix 6.5.1 of this chapter.

We use mainly SI units for computation. Conversion tables to other metric systems and the English system are given in Appendix 6.5.1.

The SI unit for the rotational speed ω is rad s^{-1}. In practical computation, however, it is often convenient to express it in terms of the stepping rate f (Hz, steps s^{-1}), and the equation of motion for this case is expressed as:

$$\tau_M = \theta_s J\frac{df}{dt} + \theta_s Df + T_f \qquad (6.4)$$

where θ_s = step angle (radian)
f = stepping rate (steps s^{-1}).

6.3.2 *Frictional torques*

The frictional torque appearing on a rotating object varies with speed as shown in Fig. 6.19. A static frictional torque T_s appears at starting, and it

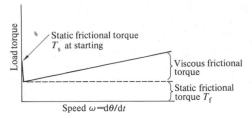

Fig. 6.19. Frictional and viscous load torque as function of speed.

at once becomes lower after the start and then increases in proportion
with the speed. The component proportional to the speed corresponds to
the second term of the right-hand side of eqn (6.3), which is due to
viscous friction. The portion indicated by T_f in the figure corresponds to
the last term in eqn 6.3. The static frictional torque T_s has no relation to
eqn (6.3) or (6.4), since those are the equations governing the motor
motion after the start. Rather T_s is an important factor which affects the
starting characteristics.

Figure 6.20 illustrates how to measure T_s and T_f. That is, a string is
wound on a pulley coupled to the load shaft, and then the loose end of the
string is pulled by a spring balance measuring forces in kg. If the force
measured just before the shaft rotates is F_s (kg) and the force just after
starting is F_f (kg), then T_s and T_f are given by

$$T_s = rF_s \quad \text{(N m)} \tag{6.5}$$

$$T_f = rF_f \quad \text{(N m)} \tag{6.6}$$

where $r =$ radius of the pulley (m).

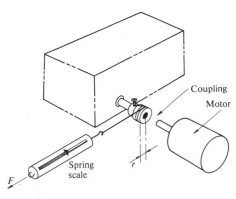

Fig. 6.20. Measuring static frictional torque.

6.3.3 *Acceleration*

We shall here discuss the relation of motor torque to acceleration rate quoting some simple cases. A motor must be started without missing steps before acceleration. But this will be discussed in the next section.

(1) *Linear acceleration.* First, when the viscous friction term is negligible, eqn (6.3) becomes

$$\tau_M - T_f = J\frac{d\omega}{dt}. \tag{6.7}$$

If the motor torque τ_M is constant in the speed range under consideration, the integration of eqn (6.3) gives

$$\omega = \{(\tau_M - T_f)/J\}t + \omega_1 \tag{6.8}$$

or the stepping rate is

$$f = \frac{\tau_M - T_f}{\theta_s J} t + f_1 \tag{6.9}$$

where ω_1 = angular speed before acceleration starts
$\quad\quad f_1$ = stepping rate before acceleration starts.
Thus the motor can be accelerated at a constant rate. This acceleration is termed 'linear acceleration'.

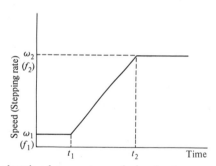

Fig. 6.21. A linear acceleration from ω_1 to ω_2 during $t_1 - t_2$.

Example 1. What is the motor torque τ_M required to accelerate an inertial load of 10^{-4} kg m^2 from $\omega_1 = 100$ to $\omega_2 = 300$ rad s^{-1} during 0.1 s, T_f being 0.05 N m?
Calculation:

$$\frac{d\omega}{dt} = \frac{\omega_2 - \omega_1}{\Delta t} = \frac{300 - 100}{0.1} = 2 \times 10^3 \text{ rad s}^{-1}; \tag{6.10}$$

$$\tau_M = J\frac{d\omega}{dt} + T_f = 10^{-4}\, 2 \times 10^3 + 0.05 = 0.25 \text{ N m}. \tag{6.11}$$

Example 2. What is the motor torque τ_M required to accelerate an inertial load of 2×10^{-4} kg m^2 from $f_1 = 500$ Hz to $f_2 = 1500$ Hz during 50 ms? Frictional load T_f is 0.03 N m? The step angle is $\theta_s = 1.18° = 3.1416 \times 10^{-2}$ rad.

Calculation:

$$\frac{df}{dt} = \frac{f_2 - f_1}{\Delta t_1} = \frac{1500 - 500}{0.05} = 2 \times 10^4; \tag{6.12}$$

$$\tau_M = \theta_s J \frac{df}{dt} + T_f$$

$$= 3.1416 \times 10^{-2} \times 2 \times 10^{-4} \times 2 \times 10^4 + 0.03 = 0.156 \quad \text{N m.} \tag{6.13}$$

Example 3. What is the maximum acceleration of an inertial load of 10^{-4} kg m^2 driven by a motor torque of 0.2 N m? Frictional loads are negligible. Step angle is $2° = 3.491 \times 10^{-2}$ rad.

Calculation:

$$\text{Acceleration} \quad \frac{d\omega}{dt} = \tau_M / J = \frac{0.2}{2 \times 10^{-4}} = 10^3 \text{ rad s}^{-2}; \tag{6.14}$$

$$\frac{df}{dt} = \frac{1}{\theta_s} \frac{d\omega}{dt} = \frac{10^3}{3.491 \times 10^{-2}} = 2.865 \times 10^4 \text{ steps s}^{-2}. \tag{6.15}$$

(2) *Exponential acceleration.* When the viscous frictional torque is not negligible the equation of motion is written as

$$\theta_s J \frac{df}{dt} + \theta_s Df - (\tau_M - T_f) = 0. \tag{6.16}$$

If the motor torque is not a function of stepping rate or speed, a solution of this differential equation is

$$f = \frac{\tau_M - T_f}{\theta_s} - \left(\frac{\tau_M - T_f}{\theta_s} - f_1\right) e^{-(D/J)t} \tag{6.17}$$

where f_1 = stepping rate at the beginning.

Thus the possible rate of acceleration decreases with stepping rate. This type of acceleration is referred to as 'exponential acceleration'.

When the motor torque τ_M is decreased with the stepping rate in a linear manner, such as

$$\tau_M = T_{M0} - \alpha f \tag{6.18}$$

the possible maximum acceleration is an exponential one as well.

(3) *Deceleration.* When a pulse train is stopped suddenly while the motor is running at a high speed, the motor may not stop at once and may

overrun. In order that a motor stops without overrunning, it may be first decelerated at a proper stepping rate from which it can safely stop.

When a stepping pulse is given at a timing a little later than the normal timing used to maintain the present speed, the generated torque τ_M will be negative. Hence the equation of motion is now

$$\theta_s J \frac{df}{dt} + \theta_s Df + (T_f - \tau_M) = 0 \qquad (6.19)$$

$$\tau_M < 0.$$

When the viscous term is negligible, the speed decreases linearly

$$f = f_s = \frac{T_f - \tau_M}{\theta_s J} t \qquad (6.20)$$

where f_s = stepping rate before deceleration is started at $t = 0$.

6.4 Determining pulse timings and intervals

We shall now develop a theory for determining pulse timings and intervals based on the preceding ideas.

6.4.1 Consideration of pull-in characteristics

It is not easy to use eqn (6.3) for the analysis of starting performance of a stepping motor, since the transient current and the rotor motion must be taken into account in computing the motor torque. Instead, the starting capability of a stepping motor is discussed using the pull-in torque characteristics. As discussed in Section 6.2.2, the starting range is a function of the load/motor inertia and varies with the method of applying the load. We must, therefore, refer to the pull-in characteristic curves for the load conditions under which the motor is to be started. Let us suppose that the pull-in torque curves show that the motor can start and stop without failure when a pulse train of f_1 Hz is provided, as illustrated in Fig. 6.22(a), the pulse interval being $1/f_1$. On computing pulse timings for acceleration for this case, it might be reasonable to set the first pulse

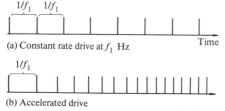

(a) Constant rate drive at f_1 Hz   Time

(b) Accelerated drive

Fig. 6.22. Constant-rate drive and accelerated drive.

interval to $1/f_1$ as in Fig. 6.22(b), and determine the subsequent pulse intervals taking account of the pull-out characteristics.

6.4.2 *Theory on pulse intervals in a linear acceleration*

We shall first discuss the linear acceleration case under the following two conditions.

(1) The stepping motor under consideration is capable of starting at the pulse rate of f_1.

(2) The motor can be accelerated at β steps s^{-2} up to the slewing rate f_s.

The first condition is related to the pull-in and the second to the pull-out characteristic.

We introduce the concept of continuously varying stepping rate f as

$$f = g + \beta t. \tag{6.21}$$

This is illustrated in Fig. 6.23 by a solid thick line. This quantity is thought of as the speed command for the motor, but the actual speed profile might be such as illustrated by the dotted curve in the same figure. The pulse timings are denoted by

$$t_1 \equiv 0, t_2, t_3, t_4 \cdots t_m \cdots .$$

Since the rotational angle instructed for each period from the mth pulse to the $(m+1)$th pulse is one step angle, the area of each trapezoid A, B, C, D $\cdots$ must be equal to one step.

Now we shall define the pulse intervals Δt_m as

$$\Delta t_m = t_{m+1} - t_m \tag{6.22}$$

and the representative pulse frequency or stepping rate f_m for the period

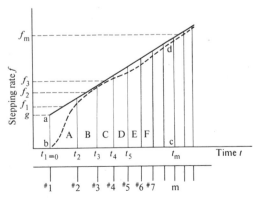

Fig. 6.23. Pulse timings in a linear acceleration.

Δt_m is defined as

$$f_m = 1/\Delta t_m. \tag{6.23}$$

The value is identical to the value of f in eqn (6.21) at $t = t_m + \Delta t_m/2$ or the midpoint of each pulse interval.

We shall now determine the quantity g in eqn (6.21) in such a way that the f at $t = \Delta t_1/2$ equals f_1. Since Δt_1, the first pulse interval, must be $1/f_1$, eqn (6.21) gives

$$f_1 = g + \beta \frac{\Delta t_1}{2} = g + \beta \frac{1}{2f_1}. \tag{6.24}$$

Consequently g is determined as

$$g = f_1 - \frac{\beta}{2f_1}. \tag{6.25}$$

The pulse timings t_m may be determined, noting that the area of the square a b c d is $(m-1)$ steps: the equation is

$$\{g + (g + \beta t_m)\} t_m = 2(m-1). \tag{6.26}$$

This gives the quadratic equation

$$\beta t_m^2 + 2g t_m - 2(m-1) = 0. \tag{6.27}$$

Hence

$$t_m = (\sqrt{(g^2 + 2(m-1)\beta)} - g)/\beta. \tag{6.28}$$

The pulse interval Δt_m is

$$\Delta t_m = t_{m+1} - t_m = \{\sqrt{(g^2 + 2m\beta)} - \sqrt{(g^2 + 2(m-1)\beta)}\}/\beta. \tag{6.29}$$

This is mathematically identical to the form

$$\Delta t_m = \frac{2}{\sqrt{(g^2 + 2m\beta)} + \sqrt{(g^2 + 2(m-1)\beta)}} \tag{6.30}$$

which yields a smaller error in numerical calculations. Hence the representative stepping rate for each pulse interval is

$$f_m = (\sqrt{(g^2 + 2m\beta)} + \sqrt{(g^2 + (2m-1)\beta)})/2 \tag{6.31}$$

Values of Δt_m and f_m for $m = 1, 2, 3 \cdots$ can be computed from eqns (6.30) and (6.31), respectively. But if f_m becomes equal to, or has exceeded, f_s at $m = M$ the pulse interval Δt_M and stepping rate f_M shall be replaced by $1/f_s$ and f_s, respectively, and the subsequent pulse intervals should be $1/f_s$. This is explained using an example in Fig. 6.24.

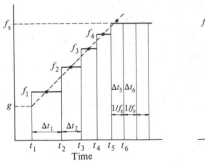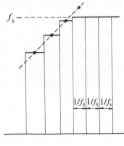

Fig. 6.24. Since f_5 exceeds f_s when computed from eqn (6.31), the values of Δt_5, Δt_6 . . . are set to $1/f_s$.

Examples. Table 6.1 shows pulse intervals computed for the following conditions

$$f_0 = 500 \text{ Hz}$$

$$f_s = 2000 \text{ Hz}$$

$$\beta = 10^5 \text{ steps s}^{-2}.$$

Twenty steps are used here before reaching slewing.

Table 6.1. Pulse timings, intervals, and rate, when started at $f_1 = 500$ Hz and accelerated at 10^5 steps s^{-2} up to $f_s = 2000$ Hz

m	t_m (ms)	Δt_m (ms)	f_m (Hz)
1	0	2.000	500
2	2.000	1.483	674
3	3.483	1.234	810
4	4.718	1.080	926
5	5.798	0.972	1028
6	6.770	0.892	1122
7	7.662	0.828	1208
8	8.490	0.776	1288
9	9.267	0.734	1363
10	10.000	0.697	1435
11	10.697	0.665	1503
12	11.362	0.638	1568
13	12.000	0.613	1631
14	12.613	0.591	1691
15	13.205	0.572	1749
16	13.776	0.554	1805
17	14.330	0.538	1860
18	14.868	0.523	1913
19	15.391	0.509	1965
20	15.900	0.500	2000
(21)	16.400		

6.4.2 *Computation to attain the slew rate at the Mth pulse*

Figure 6.25 illustrates a speed-command profile and pulse timings. In this case the speed command given by eqn (6.6) reaches the slewing rate at the Mth pulse, and the pulse intervals Δt_m for $m \geq M$ are $1/f_s$. Now we shall study a formula by which the stepping rate starts with a given initial rate f_1 and reaches a given slew rate f_s at the Mth pulse. The formulae to be derived now will also be useful for determining pulse timings and intervals for a linear deceleration.

It is found in Fig. 6.25 that f of eqn (6.21) becomes the slew rate f_s at t_M; therefore

$$f_s = g + \beta t_M. \tag{6.32}$$

From eqn (6.28)

$$\beta t_M = \sqrt{(g^2 + 2(m-1)\beta)} - g. \tag{6.33}$$

Therefore

$$f_s = \sqrt{(g^2 + 2(M-1)\beta)} \tag{6.34}$$

from which the required acceleration is obtained as

$$\beta = \frac{f_s^2 - g^2}{2(M-1)}. \tag{6.35}$$

By eliminating g from eqns (6.24) and (6.35) we obtain the quadratic equation

$$\frac{\beta^2}{4f_1^2} + (2M-3)\beta - (f_s^2 - f_1^2) = 0 \tag{6.36}$$

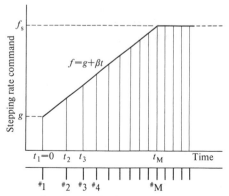

Fig. 6.25. A change from acceleration to slewing at t_M.

from which the required acceleration β is derived as follows:

$$\beta = \frac{2(f_s^2 - f_1^2)}{\sqrt{((2M-3)^2 + (f_s/f_1)^2 - 1) + (2M-3)}}. \tag{6.37}$$

The pulse timings and intervals may be computed from eqns (6.28) and (6.30), respectively, with β given by eqn (6.37) and g by eqn (6.25). But we can derive simpler equations which use f_s instead of g.

From eqn (6.34) we obtain

$$g^2 = f_s^2 - 2(M-1)\beta. \tag{6.38}$$

Substitution of this equation into eqns (6.28), (6.30), and (6.31) yields the following:

Pulse timings:

$$t_m = \frac{2(m-1)}{\sqrt{(f_s^2 - 2(M-m)\beta)} + \sqrt{(f_s^2 - 2(M-1)\beta)}}. \tag{6.39}$$

Pulse intervals:

$$\Delta t_m = \frac{2}{\sqrt{(f_s^2 - 2(M-m-1)\beta)} + \sqrt{(f_s^2 - 2(M-m)\beta)}}. \tag{6.40}$$

Stepping rates:

$$f_m = \frac{1}{\Delta t_m} = \{\sqrt{(f_s^2 - 2(M-m-1)\beta)} + \sqrt{(f_s^2 - 2(M-m)\beta)}\}/2$$

$$\text{for} \quad m = 1 \text{ to } M-1. \tag{6.41}$$

The quantities for $m \geq M$ are:

$$t_m = t_M + (m-M)/f_s \tag{6.42}$$

$$\Delta t_m = 1/f_s \tag{6.43}$$

$$f_m = f_s. \tag{6.44}$$

Example. Table 6.2 represents the pulse intervals and stepping rate for the case: $f_1 = 500$ Hz, $f_s = 2000$ Hz, $M = 20$. The acceleration β computed from eqn (6.37) is 101 075 steps s^{-2}.

6.4.3 Linear deceleration

The simplest way to determine the pulse intervals for a deceleration is to arrange in the opposite order the set of pulse intervals used for acceleration. However, since, in general, the deceleration can be higher than acceleration and the final pulse interval can be shorter than the initial one, if it is wished to use new pulse intervals for deceleration we must compute them. In that case we may apply the results of eqn (6.35) referring to Fig. 6.26.

Table 6.2. Pulse timings, intervals, and rate, starting at $f_1 = 500$ Hz and reaching the slewing rate of 2000 Hz at 20th pulse; the acceleration being 101 075 steps s^{-2}.

m	t_m (ms)	Δt_m (ms)	f_m (Hz)
1	0	2.000	500
2	2.000	1.480	676
3	3.480	1.230	813
4	4.710	1.076	929
5	5.786	0.968	1033
6	6.754	0.888	1126
7	7.642	0.824	1213
8	8.466	0.773	1294
9	9.239	0.730	1370
10	9.969	0.694	1442
11	10.663	0.662	1510
12	11.325	0.635	1576
13	11.960	0.610	1638
14	12.570	0.589	1699
15	13.159	0.569	1758
16	13.728	0.551	1814
17	14.279	0.535	1869
18	14.814	0.520	1923
19	15.334	0.506	1974
$M = 20$	15.840	0.500	2000
(21)	16.340		

The rate of deceleration γ which uses N pulses in deceleration and makes the final pulse interval equal $1/f_l$ is given by

$$\gamma = \frac{2(f_s^2 - f_1^2)}{\sqrt{((2N-1)^2 + (f_s/f_1)^2 - 1)} + (2N-1)} \tag{6.45}$$

where $f_1 =$ final stepping rate
$f_s =$ slewing stepping rate.

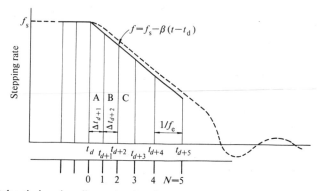

Fig. 6.26. Pulse timings in a linear deceleration.

Using this γ, the pulse intervals and pulse rates are calculated from the following equations.

$$\Delta t_{d+n} = 2/\{\sqrt{(f_s^2 - 2n\gamma)} + \sqrt{(f_s^2 - 2(n-1)\gamma)}\} \qquad (6.46)$$

$$f_{d+n} = \frac{1}{\Delta t_{d+n}} \{\sqrt{(f_s^2 - 2n\gamma)} + \sqrt{(f_s^2 - 2(n-1)\gamma)}\}/2 \qquad (6.47)$$

where d is the final pulse number in the slewing region.

Example. Table 6.3 shows the pulse interval and stepping rate computed under the following conditions:

$$f_s = 2000 \text{ Hz}$$
$$f_1 = 600 \text{ Hz}$$
$$N = 15.$$

The deceleration computed from eqn (6.45) is

$$\gamma = 125\,142 \text{ steps s}^{-2}.$$

6.4.4 Exponential acceleration

In most cases the pull-out torque decreases with stepping rate. If it is assumed that the pull-out torque is the upper limit of torque necessary to

Table 6.3. Example of pulse intervals in a linear deceleration starting from 2000 Hz and slowing to 600 Hz with 15 pulses.

m	Δt_{d+n} (ms)	f_{d+n} (Hz)
0	0.500	2000
1	0.508	1968
2	0.525	1904
3	0.544	1837
4	0.566	1767
5	0.590	1695
6	0.618	1619
7	0.649	1540
8	0.687	1456
9	0.731	1368
10	0.786	1273
11	0.855	1170
12	0.946	1057
13	1.074	931
14	1.275	784
15	1.667	600

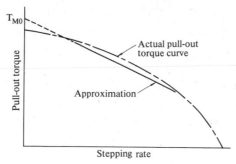

Fig. 6.27. A linear approximation of pull-out torque.

carry and accelerate the load, the motion after the start must satisfy the relation.

$$J\theta_s \frac{df}{dt} + D\theta_s f + T_0 < \tau_{\text{pull-out}}. \tag{6.48}$$

Let us approximate the torque created by a motor by

$$\tau_M = T_{M0} - \alpha f \tag{6.49}$$

in the working range of speed as shown in Fig. 6.27, and discuss the pulse intervals for this case. Now the equation of motion is:

$$J\theta_s \frac{df}{dt} + D\theta_s f + T_0 = T_{M0} - \alpha f. \tag{6.50}$$

Equation (6.50) is rewritten in the form

$$J\theta_s \frac{df}{dt} + (D\theta_s + \alpha)f - (T_{M0} - T_0) = 0. \tag{6.51}$$

The solution of this equation is

$$f = g + \left(\frac{T_{M0} - T_0}{D\theta_s + \alpha} - g \right) \left\{ 1 - \exp \left(-\frac{D\theta_s + \alpha}{J\theta_s} t \right) \right\}. \tag{6.52}$$

The form of f as a function of t is illustrated in Fig. 6.28. (Pulse timings are also indicated in the same figure.) Here g is the value of f at $t = 0$ and it should be determined such that the first pulse interval becomes $1/f_1$ as in the case of the linear acceleration. Here, f_1 is a stepping rate at which the motor can start without failure.

The integration of eqn (6.52) is the rotational angle X expressed in steps:

$$X = \int_0^t f \, dt = \left(\frac{T_{M0} - T_0}{K} \right) t + \left(\frac{T_{M0} - T_0}{K} - g \right) \left(\frac{J\theta_s}{K} \right) \left\{ \exp \left(-\frac{K}{J\theta_s} t \right) - 1 \right\} \tag{6.53}$$

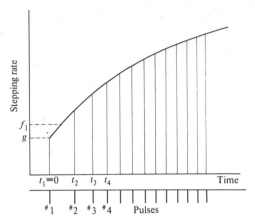

Fig. 6.28. Pulse timings in an exponential acceleration.

where

$$K = \alpha + \theta_s D. \tag{6.54}$$

Since X must be 1 step at $t = 1/f_1$, g is determined as

$$g = \frac{T_{M0} - T_0}{K} - \left(1 - \frac{T_{M0} - T_0}{f_1 K}\right)\left(\frac{K}{J\theta_s}\right) \bigg/ \left\{\exp\left(\frac{-K}{J\theta_s f_1}\right) - 1\right\}. \tag{6.55}$$

Hence we obtain the expression for X as follows:

$$X(t) = \left(\frac{T_{M0} - T_0}{K}\right)t - \left(\frac{T_{M0} - T_0}{f_1 K} - 1\right)\frac{\exp\left(-Kt/J\theta_s\right) - 1}{\exp\left(-K/J\theta_s f_1\right) - 1}. \tag{6.56}$$

The pulse timings can be computed by solving the equation

$$X(t_m) - m = 0 \tag{6.57}$$

where $m = 1, 2, 3, 4 \cdots$

The pulse intervals and rate are respectively:

$$\Delta t_m = t_{m+1} - t_m \tag{6.58}$$

$$f_m = 1/\Delta t_m. \tag{6.59}$$

These can be determined by a computer using a small program. The initial rate of acceleration is defined as

$$\beta_{\text{initial}} = \left(\frac{\partial f}{\partial t}\right)_{t=0} = \{(T_{M0} - T_0)/K - g\}(K/J\theta_s)\exp\left(-K/J\theta_s f_1\right). \tag{6.60}$$

Table 6.4. Example of pulse intervals in
an exponential acceleration.

m	t_m (ms)	Δt_m (ms)	f_m (Hz)
1	0	2.000	500
2	2.000	1.495	669
3	3.495	1.257	796
4	4.752	1.109	902
5	5.862	1.007	993
6	6.869	0.930	1076
7	7.798	0.870	1150
8	8.668	0.821	1218
9	9.489	0.781	1281
10	10.270	0.746	1341
11	11.016	0.716	1396
12	11.732	0.690	1448
13	12.423	0.668	1498
14	13.090	0.647	1545
15	13.737	0.629	1590
16	14.366	0.612	1633
17	14.978	0.597	1675
18	15.575	0.583	1715
19	16.159	0.570	1753
20	16.729	0.559	1790
21	17.287	0.548	1826
22	17.835	0.537	1861
23	18.373	0.528	1894
24	18.901	0.519	1926
25	19.420	0.511	1958
26	19.930	0.503	1988
27	20.433	0.496	2018
28	20.929		

Table 6.4 shows an example of computed results under the following
conditions:

Maximum torque	$T_{M0} = 0.4\,\text{N m}$
Static friction	$T_0 = 0.05\,\text{N m}$
Slope of approximated torque line	$\alpha = 5 \times 10^{-5}\,\text{N m s step}^{-1}$
Inertia	$J = 10^{-4}\,\text{kg m}^2$
Step angle	$\theta_s = 1.8\,\text{degrees}\;(=0.031416\,\text{rad})$
Viscosity	$D = 10^{-3}\,\text{N m s rad}^{-1}$
Starting rate	$f_1 = 500\,\text{Hz}\;(=\text{steps s}^{-1})$

The initial acceleration for this case is:

$$\beta_{\text{initial}} = 95\;921.2 \text{ steps s}^{-2}.$$

6.5 Appendixes

6.5.1 *Appendix—equivalent inertia at the motor shaft*

(1) *Reflected inertia through belt or gears.* When the motor torque is transmitted through gears or a belt and pulleys as shown in Fig. 6.29 the overall inertia reflected at the motor shaft is

$$J = \left(\frac{Z_1}{Z_2}\right)^2 J_2 + J_1 \qquad (6.61)$$

or

$$J = \left(\frac{D_1}{D_2}\right)^2 J_2 + J_1 \qquad (6.62)$$

where J = total equivalent inertia at the motor shaft
  J_1 = inertia of gear 1 or pulley 1
  J_2 = inertia at the load shaft; of the load, shaft, and gear 2 or pulley 2
  Z_1 = number of teeth in gear 1
  Z_2 = number of teeth in gear 2
  D_1 = diameter of pulley 1
  D_2 = diameter of pulley 2.

(2) *Lifting up an object* (Fig. 6.30). When a motor lifts an object of mass M by means of a pulley of inertia J_1, the overall reflected inertia at the motor is

$$J = J_1 + \tfrac{1}{4}MD^2. \qquad (6.63)$$

(3) *Driving an object by means of a belt* (Fig. 6.31)

$$J = 2J_1 + \tfrac{1}{4}MD^2 \qquad (6.64)$$

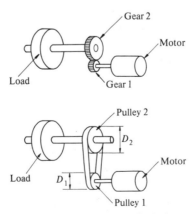

Fig. 6.29. Torque transmission by gears or pulleys and a belt.

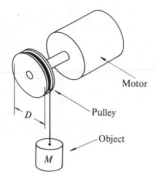

Fig. 6.30. Lifting an object by means of a belt.

where J_1 = inertia of each pulley (kg m²)
D = pulley diameter (m)
M = mass of the object and belt (kg).

(4) *Linear motion by means of a lead screw and gears.* When a workpiece
and a table are driven by means of gears and a lead screw as shown in
Fig. 6.32, the overall inertia reflected at the motor is

$$J = J_1 + \left(\frac{Z_1}{Z_2}\right)^2 (J_2 + J_3) + M\left(\frac{p}{2\pi}\frac{Z_1}{Z_2}\right)^2 \qquad (6.65)$$

where J_1 = inertia of gear coupled to the rotor (kg m²)
J_2 = inertia of gear coupled to the lead screw (kg m²)
J_3 = inertia of the lead screw (kg m²)
Z_1 = number of teeth on the gear coupled to the rotor
Z_2 = number of teeth on the gear coupled to the lead screw
M = mass of the object and table (kg)
p = pitch of the lead screw (m)

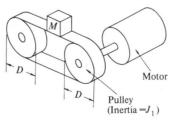

Fig. 6.31. Linear movement of an object by means of a belt.

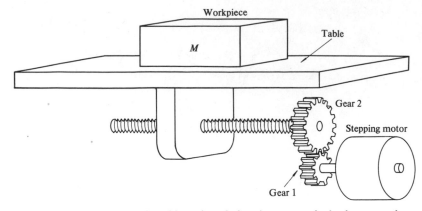

Fig. 6.32. Linear movement of a table and workpiece by means of a lead screw and gears.

6.5.2 Appendix—tables of unit conversions

(1) Weight/Mass

	Kilogramme (kg)	Pound (lb)	Ounce (oz)
kg	1	2.204 62	35.2739
lb	0.453 592	1	16
oz	0.028 349	0.0625	1

(2) Length

	Metre (m)	Inch (in)	Foot (ft)
m	1	39.3707	3.280 89
in	0.025 399	1	0.083 33
ft	0.304 794	12	1

(3) Torque

	Newton metre (N m)	Pound inch (lb in)	Ounce inch (oz in)
N m	1	8.850 75	141.612
lb in	0.112 985	1	16
oz in	0.007 061 55	0.0625	1

(4) *Rotational speed*

	Radian per second (rad s^{-1})	Revolutions per second (r.p.s.)	Revolutions per minute (r.p.m.)
rad s^{-1}	1	0.159 155	9.549 29
r.p.s.	6.283 19	1	60
r.p.m.	0.104 719	0.016 6667	1

(5) *Moment of inertia*

	kg m^2	oz in s^2	lb in s^2
kg m^2	1	141.612	8.850 73
oz in s^2	0.007 061 55	1	0.0625
lb in s^2	0.112 985	16	1

References for Chapter 6

[1] Kenjo, T. and Niimura, Y. (1979). Fundamentals and applications of stepping motors. (In Japanese.) pp. 225–7. Sogo Electronics Publishing Co., Ltd., Tokyo.
[2] Ward, P. A. and Lawrenson, P. J. (1977). Backlash, resonance and instability in stepping motors. *Proc. Sixth annual symposium on Incremental motion control systems and devices*. Department of Electrical Engineering, University of Illinois, pp. 73–83.
[3] Kordic, K. S. (1976). Dynamic torque measurements for step motors. *Proc. Fifth Annual Symposium on Incremental motion control systems and devices*. Department of Electrical Engineering, University of Illinois, pp. E1–21.

7. Closed-loop control of stepping motors

7.1 Limitations of open-loop operation and need for closed-loop operation

In the drive systems explained in Chapter 5 the step-command pulses were given from an external source, and it was expected that the stepping motor is able to follow every pulse. This type of operation is referred to as the open-loop drive. The open-loop drive is attractive and widely accepted in applications of speed and position controls. However, the performance of a stepping motor is limited under the open-loop drive. For instance a stepping motor driven in the open-loop mode may fail to follow a pulse command when the frequency of the pulse train is too high or the inertial load is too heavy. Moreover the motor motion tends to be oscillatory in open-loop drives.

The performance of a stepping motor can be improved to a great extent by employing position feedback and/or speed feedback to determine the proper phase(s) to be switched at proper timings. This type of control is termed the closed-loop drive. In a closed-loop control, a position sensor is needed for detecting the rotor position. As a typical sensor, nowadays, an optical encoder is used and it is usually coupled to the motor shaft. The mechanism and principle of an optical encoder will be explained later using Figs. 7.10 and 7.11. In a more advanced system, instead of an additional mechanical sensor, rotor position is sensed by observation of waveforms of the currents in the motor windings.

The closed-loop control is advantageous over the open-loop control not only in that step failure never occurs but that the motion is much quicker and smoother.

7.2 The concept of lead angle

Before discussing the details of the closed-loop system we shall study the concept of lead angle.

7.2.1 *One-step lead angle and bigger lead angles*

Let us suppose that in the closed-loop system of Fig. 7.1 a stepping motor is running or about to start. The optical encoder coupled to the rotor detects the rotor position and supplies its information to the logic sequencer. Then, the logic sequencer determines the proper phase(s) to be excited, taking account of the position information. The relation between the rotor's present position and the phase(s) to be excited is

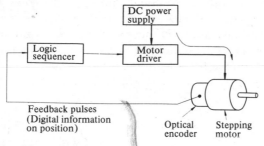

Fig. 7.1. Simple closed-loop operation of a stepping motor.

specified in terms of lead angle. In this example the motor is a three-phase motor and the sequence of excitation is Ph1 → Ph2 → Ph3 in the single-phase-on mode. Ph1 is now excited and the rotor is stopping at an equilibrium position. Then Ph2 is excited and Ph1 is de-energized to start the motor. The lead angle in this case is one step. As soon as the positional encoder detects that the rotor reaches an equilibrium position of $Ph(N)$, the logic sequencer set for operation of one-step lead angle will generate the signal to turn on $Ph(N+1)$ to continue the motion. Thus a stepping motor in a closed-loop system runs like a brushless DC motor in which the proper winding(s) to be energized is(are) automatically selected by a position sensor incorporated in or coupled to the motor.

The speed of a stepping motor driven in a closed-loop mode varies with load as shown in Fig. 7.2. The bigger the load, the slower the speed. However, a lead angle of one step is not *always* used, since it *may* not ensure a continuous rotation when a frictional load is carried. The reason is as follows. Suppose that a motor is proceeding towards an equilibrium position of the phase now excited. Since the static torque decreases as the rotor comes closer to the equilibrium position, the motor may, before reaching the equilibrium position, stop at the position where the static

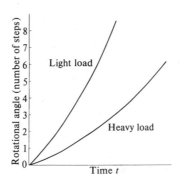

Fig. 7.2. Travelled distance vs. time characteristics in closed-loop operation.

torque and friction torque are equal and opposite. Now since the lead angle is one step, the next phase is not turned on, which means that the rotor can not run any more. If the switching of excitation is done at a proper position before the equilibrium position, the motor can continue to run. This is the case for a lead angle larger than one step.

7.2.2 Lead angle and static torque

The torque vs. displacement curves for a three-phase motor can be approximated by the sinusoidal waves shown in Fig. 7.3(a). When the motor is running in the one-phase-on mode with a one-step lead angle, the static torque will vary with time in the manner traced by the thicker curves in Fig. 7.3(b). Note that when the rotor reaches an equilibrium

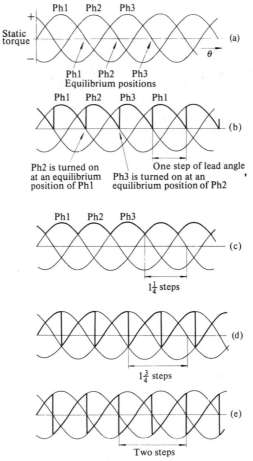

Fig. 7.3. Relation between lead angle and static torque in a three-phase motor: (a) characteristics for each phase; (b)–(e) for various lead angles from 1 to 2 steps.

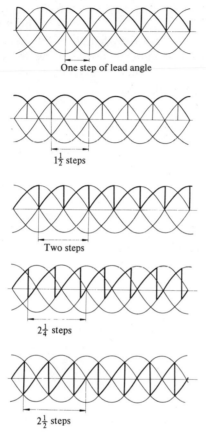

One step of lead angle

$1\frac{1}{2}$ steps

Two steps

$2\frac{1}{4}$ steps

$2\frac{1}{2}$ steps

Fig. 7.4. Relation between lead angle and static torque in a four-phase motor.

position of the phase so far excited, this phase is turned off and the next phase is turned on. Figure 7.3(c) shows the case of $1\frac{1}{4}$-step lead angle which gives the highest mean torque in the sinusoidal model. If the lead angle is further increased, the mean static torque will decrease (see (d) and (e)).

The case of a four-phase motor is shown in Fig. 7.4, which shows that the mean torque becomes highest when the lead angle is 1.5 steps.

7.2.3 Big lead angles for high speeds

Again in Fig. 7.3(e), it is seen that the mean static torque developed in a three-phase motor is zero if the lead angle is set to 2.0 steps. Likewise, a four-phase motor with lead angle set to 2.5 steps produces zero mean static torque. These statements are, however, true only for the case when the speed is extremely low. When a motor is running, those lead angles produce mean torques sufficient to accelerate or maintain speed. This is due to the electrical time constant of windings. The torque–displacement curves in Figs. 7.3 and 7.4 are drawn under the assumption that current builds up to its maximum value as soon as the transistor is turned on. Actually however, there is a time delay before the current reaches its maximum due to the winding inductance.

In Fig. 7.5 the relationship between the voltage and current is shown for three different stepping rates: (a) a low speed; (b) a medium speed; and (c) a high speed. The current flowing through the windings in the transistor's ON period produces positive torque useful for accelerating the motor. On the other hand, the current which circulates in the winding and suppressor diode after the transistor's turning-off can be the cause of retarding torque. When the speed is low, the effect of the retarding torque is negligible. As speed increases, however, as is obvious in Fig. 7.5(b) and (c), the period for positive torque becomes comparably small and the average torque will be decreased. Hence the maximum torque is not so high as long as the lead angle is set to a value close to 1.0 step. The mean torque and maximum speed are increased by increasing the lead angle, since the time lag in current build-up is offset by earlier turning-on as shown in Fig. 7.6. Shimotani and Kataoka[1] made a theoretical analysis of this matter for a multi-stack VR motor.

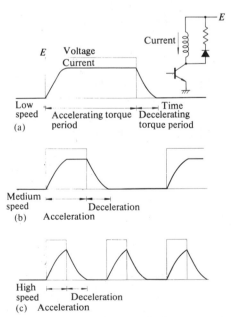

Fig. 7.5. Relation between the voltage and current in the one-step lead angle operation.

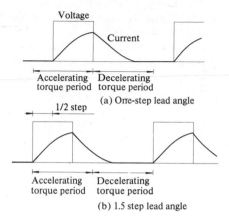

(a) One-step lead angle

(b) 1.5 step lead angle

Fig. 7.6. Mean torque is increased by increasing lead angle in higher speed ranges.

Figure 7.7[2] illustrates an example of maximum slew speed vs. lead angle characteristics which was measured on a four-phase hybrid stepping motor driven in the two-phase-on mode. In this case the maximum slew speed is about 1000 Hz with 1.0-step lead angle, but it is increased to 14 000 Hz with 3.2-step lead angle. Figure 7.8 shows another example, in which the lead angle is varied not continuously, but in a discrete manner from 1.0 to 3.0 steps by increments of 0.5 step. Single-phase excitation is used with 1.5 and 2.5-step lead angles, while two-phase excitation is used for 1.0, 2.0, and 3.0-step lead angles. The maximum speed is over 25 kHz

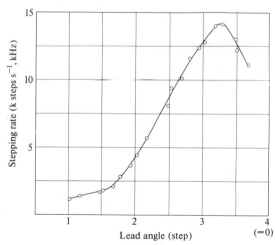

Fig. 7.7. Maximum slew speed vs. lead angle measured on a 1.8° hybrid stepping motor driven in the two-phase-on mode.

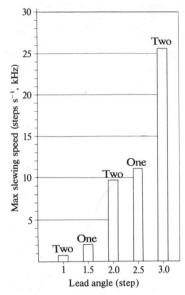

Fig. 7.8. Maximum slewing speed vs. lead angle measured on a 1.8° hybrid motor in the combined mode of one- and two-phase excitation.

when the lead angle is set to 3.0 steps. The engineering meanings of the latter case will be discussed in Section 7.2.5.

7.2.4 *Definition of lead angles for two-phase-on operation*

The discussion of the lead angle has so far been concentrated on the single-phase-on drive. Lead angles can be defined for the two-phase-on

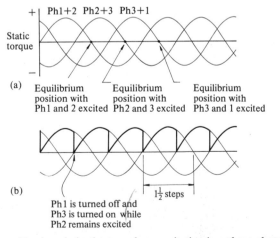

Fig. 7.9. Definition of lead angle in the two-phase excitation in a three-phase motor.

mode, too. Figure 7.9(a) shows the static torque curves as functions of displacement (= angular position), approximated by sinusoidal waves, when two of the three phases are excited. Some equilibrium positions are indicated in the same figure. When the rotor arrives at an equilibrium position, with Ph1 and Ph2 excited, Ph1 is turned off and Ph3 is turned on, this is the drive of one-step lead angle. The static torque curve in Fig. 7.9(b) is the case of 1.5-step lead angle.

7.2.5   Combination of single-phase and two-phase excitation

In most actual cases of closed-loop operation, an optical encoder or another type of position sensor is coupled to the rotor to detect rotor position. Unless the number of encoder output pulses per revolution is many more than the steps per revolution, it is difficult to vary the lead angle continuously. In many applications, possible lead angles are limited. For example, if the encoder resolution is 200 pulses per revolution and the number of steps of a stepping motor per revolution is also 200, a possible set of lead angles will be 0, 1, 2, and 3 steps. Another possible choice is a set of 0.5, 1.5, 2.5, and 3.5 steps. Which of the first or second set occurs depends on the rotor/encoder coupling, and on choice of single-phase or two-phase drive. If both single-phase and two-phase operations are employed, however, the system can use eight different lead angles.

This idea was suggested by Fredriksen in a paper[3] in 1967. The position encoder shown in his paper is an assembly of an opaque disc with 50 small holes 7.2° apart and four pairs of photo-transistor/light sources. Nowadays, multichannel optical encoders are widely used for this purpose. Discussions on the lead angle in the combined mode of the single-phase-on and two-phase-on operation are here made using a three-channel optical encoder of resolution 200 pulses per revolution. Figure 7.10 shows a model of an optical encoder directly coupled to the shaft, where only one channel is shown for simplicity.

The principle of the optical encoder is illustrated in Fig. 7.11. It is constructed with a light source, a sensor, a rotary disc, and a stationary mask. The disc has alternate opaque and transparent sectors. As the disc rotates with the motor, the mask passes and periodically blocks the light. The output signal from the sensor is formed to be suitable as a digital signal.

One of the possible relations between the encoder output signal and equilibrium positions of the single-phase excitation mode is illustrated in Fig. 7.12. The encoder is coupled to the rotor such that the equilibrium positions occur at the centres of the H-level intervals in the A-channel output. The A-channel is used for discrimination of the rotational direction as explained later. The B-channel signal, shifted by 90° compared

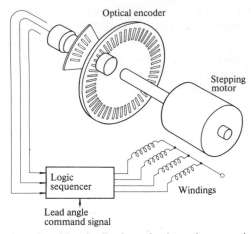

Fig. 7.10. Configuration of position feedback mechanism using an optical encoder.

with the A-channel signal, is employed for generation of clock pulses which are used as switching signals. Two clock generators, CG1 and CG2, are used here. CG1 shoots a negative logic clock pulse whenever the B-channel signal decreases in an H-level interval of the A-channel signal. CG2 shoots a pulse at a build-up of the B-channel signal in an L-level

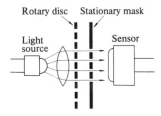

Fig. 7.11. Principle of the optical encoder.

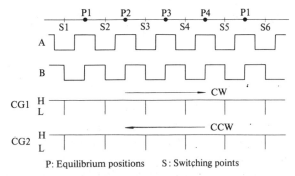

P: Equilibrium positions   S: Switching points

Fig. 7.12. Relation between the encoder's output signals and single-phase-on equilibrium positions.

interval of the A-channel signal. In other words, CG1 produces clock pulses when the motion is clockwise (CW) while CG2 works when the motion is counter-clockwise (CCW). The rotor positions at which clock pulses are generated are here termed switching points.

A relation between the possible lead angles, the one-phase-on equilibrium positions, and the phases to be excited is given in Table 7.1 for both directions. Let us follow, as an example, the row of 1.5-step lead angle. When the rotor crosses the switching point S2, Ph2 is turned off and Ph3 is turned on. The distance from S2 to the next equilibrium position for Ph3 excited is 1.5 steps. If Ph4 is turned on at S2, the lead

Table 7.1. Relation of lead angles, one-phase-on equilibrium positions, and phases to be excited in a four-phase motor under the arrangement that switching is done at two-phase-on equilibrium positions.

CW Lead angle	S1 P1	S2 P2	S3 P3	S4 P4	S5 P1	S6 P2		
0	4, 1	1, 2	2, 3	3, 4	4, 1		#1	
0.5	1	2	3	4	1		#2	Stop, CW
1.0	1, 2	2, 3	3, 4	4, 1	1, 2		#3	
1.5	2	3	4	1	2		#4	Revs, CW
2.0	2, 3	3, 4	4, 1	1, 2	2, 3		#5	Accel, CW
2.5	3	4	1	2	3		#6	
3.0	3, 4	4, 1	1, 2	2, 3	3, 4		#7	
3.5 (= −0.5)	4	1	2	3	4		#8	Decel, CW

CCW lead angle	S1 P1	S2 P2	S3 P3	S4 P4	S5 P1	S6 P2		
0	1, 2	2, 3	3, 4	4, 1	1, 2		#1	
0.5	1	2	3	4	1		#2	Stop, CCW
1.0	4, 1	1, 2	2, 3	3, 4	4, 1		#3	
1.5	4	1	2	3	4		#4	Revs, CCW
2.0	3, 4	4, 1	1, 2	2, 3	3, 4		#5	Accel, CCW
2.5	3	4	1	2	3		#6	
3.0	2, 3	3, 4	4, 1	1, 2	2, 3		#7	
3.5 (= −0.5)	2	3	4	1	2		#8	Decel, CCW

Notes (1) P1, P2, P3, and P4 are one-phase-on equilibrium positions.
    (2) S1, S2 · · · are switching points in this arrangement.

#1: Effective braking; does not ensure reversing
#2: Poor braking; used for final positioning
#3: Not good for running
#4: Good starting, slow slewing
#5: Good starting/acceleration
#6: Does not start motor, but provides high slew speed
#7: Starts motor in opposite direction, provides highest slewing in normal direction
#8: Good braking

Table 7.2. Relation between lead angles, equilibrium positions, and phases to be excited in a four-phase motor under the arrangement that switching is effected at the single-phase-on equilibrium positions.

CW lead angle	S200		S1		S2		S3		S4
	P4	P_{41}	P1	P_{12}	P2	P_{23}	P3	P_{34}	P4
0		4		1		2		3	
0.5		4, 1		1, 2		2, 3		3, 4	
1.0		1		2		3		4	
1.5		1, 2		2, 3		3, 4		4, 1	
2.0		2		3		4		1	
2.5		2, 3		3, 4		4, 1		1, 2	
3.0		3		4		1		2	
3.5 (= −0.5)		3, 4		4, 1		1, 2		2, 3	

CCW lead angle	S200		S1		S2		S3		S4
	P4	P_{41}	P1	P_{12}	P2	P_{23}	P3	P_{34}	P4
0		1		2		3		4	
0.5		4, 1		1, 2		2, 3		3, 4	
1.0		4		1		2		3	
1.5		3, 4		4, 1		1, 2		2, 3	
2.0		3		4		1		2	
2.5		2, 3		3, 4		4, 1		1, 2	
3.0		2		3		4		1	
3.5 (= −0.5)		1, 2		2, 3		3, 4		4, 1	

Notes (1) P1, P2, P3, and P4 are the single-phase-on equilibrium positions.
  (2) P_{12}, P_{23}, P_{34}, and P_{41} are the two-phase-on equilibrium positions.
  (3) S1, S2 $\cdots$ are switching points in this arrangement, being different from S1, S2 $\cdots$ in Table 7.1.

angle is 2.5 steps, as occurring in the sixth row. Thus the one-phase excitation covers 0.5, 1.5, 2.5, and 3.5 steps of lead angle. The lead angle of 3.5 steps may be regarded as −0.5-step lead angle. If Ph2 and Ph3 are turned on at S2 in the CW direction as occurring in the third row, the lead angle is 1.0 step, since the next equilibrium position where Ph2 and Ph3 are equally excited is at the switching point S3 which is 1.0 step away from S2. The two-phase excitation covers 0, 1.0, 2.0 and 3.0 steps of lead angle.

In Table 7.1 the switching points occur at the two-phase-on equilibrium positions. If the encoder is coupled in such a way that the switching points occur at the single-phase-on equilibrium positions, the relation between the lead angles and the phases to be excited is as shown in Table 7.2.

7.2.6 *Starting lead angle and running lead angle*

When a system is so designed as to position the rotor at an intermediate point between two successive switching points, the lead angle at stopping or starting is less than that at running by 0.5 step. This occurs, for example, in Table 7.1 if the final positioning is made in the single-phase-on mode. In order to position the rotor at P1, only Ph1 should be excited, and likewise, Ph2 should be excited to accommodate the rotor at an equilibrium position P2. These correspond to the second row in Table 7.1 both for CW and CCW directions. When the motor is stopping in this mode, the lead angle is 0 step. But, if the motor is moving in either direction, the lead angle is 0.5 step. The difference of 0.5 step is seen also in the other modes. Suppose the motor is initially positioned at P1 with Ph1 excited. The motor is, then, started in the CW direction in the mode governed by the fifth row, say with a running lead angle of 2.0 steps. The phases now excited are Ph2 and Ph3, and the equilibrium position with this excitation is position S3 which is 1.5 steps away from P1. The starting lead angle is thus less than 2.0 steps by 0.5 step.

This difference of lead angles between starting and running has the advantage that a four-phase motor can be started in the desired direction in the 2.0-step lead angle mode. If the lead angle is also set to the normal 2.0 step mode at starting, the direction of rotation is not known.

7.2.7   Lead angles for acceleration and deceleration

When a stopping motor is to be accelerated, the lead angle must be larger than 1.0 step. Figure 7.13 shows some examples of speed against distance characteristics for three different lead angles, 1.0, 1.5, and 2.0 steps. Those were the curves measured on the same motor/driver combination that were used to get the maximum slew speed characteristics in Fig. 7.8. It is obvious that the 1.0-step lead angle accelerates the motor less effectively, while the 2.0-step lead angle provides a good acceleration. If the lead angle is changed to a bigger value after the motor has been

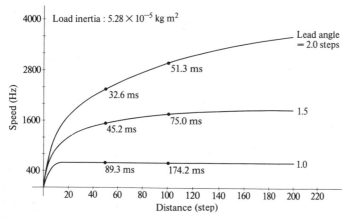

Fig. 7.13. Plots of speed against number of steps based on Table 7.1, with lead angles of 1.0, 1.5, and 2.0 steps (for acceleration).

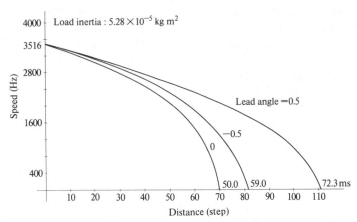

Fig. 7.14. Plots of speed against number of steps based on Table 7.1, with lead angles of −0.5, 0, and 0.5 steps (for deceleration).

accelerated to a certain speed with 2.0-step lead angle, the acceleration is more effective.

To decelerate the rotor, the lead angle should be set to zero or to some small value positive or negative. Figure 7.14 shows plots of speed against number of steps with lead angles of −0.5, 0, and 0.5 steps.

7.3   A closed-loop operation system using a microprocessor

In Fredriksen's paper[3] quoted before, he suggested the application of a big or minicomputer to the closed-loop control of stepping motors. Since nowadays, microprocessors are available cheaply, utilization of a microprocessor in control of stepping motors is a very interesting engineering problem. This section discusses a microprocessor-controlled closed-loop system designed by Kenjo et al.[4] for the drive of a 1.8° four-phase hybrid stepping motor. The microprocessor used is 8080A type. The choice of lead angles and the arrangement of switching points are based on Table 7.1. The positional detector used is a three-channel optical encoder. Two channels, A and B, are used for the same function as described in Section 7.2.5. The third channel, R, produces one pulse per revolution, and is used for detecting the reference position.

7.3.1   Purpose of the microprocessor

Figure 7.15 shows two curves of speed versus distance plots under closed-loop control of a step motor. The (a) curve is an excellent pattern where the motor is started with an appropriate lead angle, accelerated with another lead angle, and begun to be decelerated at the best timing

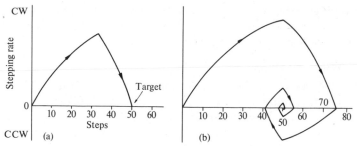

Fig. 7.15. Two curves of speed vs. distance plots: (a) excellent curve; (b) unskilled curve.

from which the speed is reduced most quickly and becomes zero just at the target. To start and accelerate a motor, a lead angle larger than one step is used, while zero or a negative lead angle is used for deceleration. On curve (b), which is an unskilled one, deceleration is initiated when the target position is detected. But the motor cannot stop at once and will overrun due to inertia. To accommodate the rotor at the correct position, the motor is forced to move backward by setting the lead angle to a proper value. The speed versus distance locus may be oscillatory as shown in the figure.

A microprocessor system is used here to find out the best timings to change lead angles to attain the (a) type motion, starting from the pattern of (b). Figure 7.16 illustrates the outline of the system, which has a dedicated logic sequencer outside the microprocessor. A positional signal is fed back to the block of hardware which monitors the rotor movement and exchanges information with the microprocessor. The software must be programmed so that the microprocessor determines

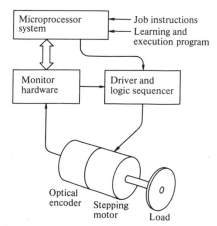

Fig. 7.16. System outline.

better timings for changing lead angles, based on the previous experience and present position/speed data. The microprocessor will finally, after several executions, find the optimum timings for each motion used.

7.3.2 *System details and hardware*

The control system is shown in a block diagram in Fig. 7.17. When an instruction, for example to print a character or to position at a particular point, is given by a supervisory computer or another system, the microprocessor calculates the distance to the target and direction of rotation. The microprocessor is also used for determining timings to set proper lead angles, judging the following items of information:
  (i) present position;
 (ii) error steps from target;
(iii) signal to indicate that the rotor has passed the point a half-step before target;
(iv) signal to indicate that motion has reversed;
 (v) speed.
The last item is not used in the example described here.

(1) *Clock generator/direction discriminator.* This block receives the three-channel signal from the position encoder, and generates the clock pulses used to operate the hardware outside the microprocessor. The relation between the encoder signals and the pulse timings has already been explained in 7.2.5, using Fig. 7.12. A direction discriminator which is a set–reset flip–flop shown in Fig. 7.18 is incorporated in this block. The clock signals from CG1, which are shot when clockwise, are fed to the $\bar{S}$ terminal, and the clock signals from CG2, which are shot when counterclockwise, are fed to the $\bar{R}$ terminal. If a clock pulse is shot by CG1, the state of Q1 will become H. More pulses from CG1 do not change the states of Q1 and Q2. In the situation shown in Fig. 7.19, where the motor has moved up to near an equilibrium position P3 in the CW direction and is about to reverse, the Q1 terminal of the flip–flop remains at H-level until the switching point S3 is crossed in the CCW direction. When the motor moves across the S3 point, a clock pulse is generated by CG2 to change the output states to Q1 = L and Q2 = H. There is the possibility that the rotor oscillates about an equilibrium position as shown in Fig. 7.20. But the direction discriminator does not change unless any switching points are crossed.

(2) *Phase counter.* This block discriminates which of the four zones P1, P2, P3, or P4 the motor is passing or stopping at.

(3) *Logic sequencer.* This block determines the phases to be excited, receiving instructions for lead angle and direction from the microprocessor, and referring to the information from the phase counter. This

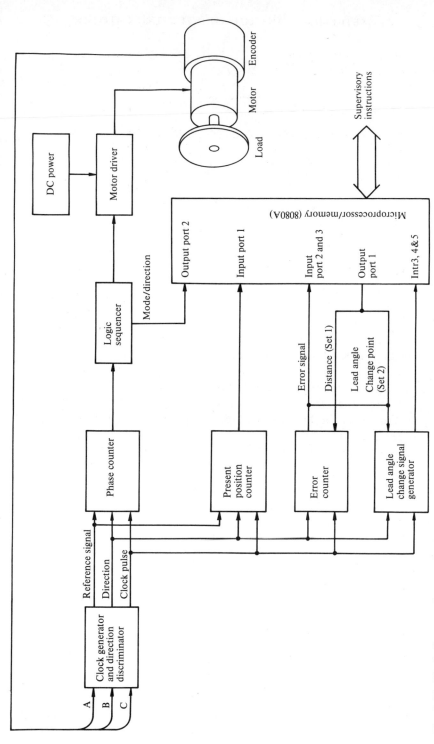

Fig. 7.17. Block diagram of system hardware.

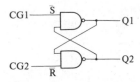

Fig. 7.18. Set–reset flip–flop used for a direction discriminator.

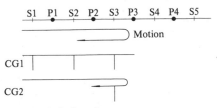

Fig. 7.19. Reverse of motion and clock pulses.

rule is given in Table 7.1. The characters of each mode of lead angle are summarized in the notes to the table.

(4) *Stop mode of lead angle and final positioning.* The lead angle is set to 0.5 step or stop mode by hardware whenever the motor is in the target zone which is the one-step interval centring on the target. If the motor speed is slow enough when the rotor has entered the target zone, it may be decelerated and eventually stops at the target equilibrium point. In this system, therefore, the final positioning is carried out by the single-phase excitation. If the rotor passes the target zone and fails to stop, the lead angle is set by software. As will be stated, the lead angle is also set by software before entering the target zone.

(5) *Motor driver.* The power circuit to drive the motor may be any type discussed in Section 4.3.

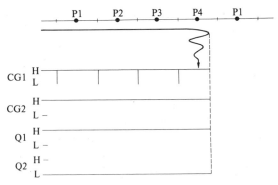

Fig. 7.20. Small oscillation about an equilibrium position does not change the state of the direction discriminator.

(6) *Present position counter.* This is a counter to record the present position. When the reference position is detected while rotating in the CW direction, the counter is reset to zero and counts up whenever it receives a clock pulse from CG1. When the reference position is detected while in CCW rotation, the counter is set to 200 in order to make the next count 199. The counter counts down in the CCW direction.

(7) *Error counter.* This block counts the distance between the present position and the target position, and sends this item of information to Input Ports 2 and 3 in the microprocessor. The target position is instructed by the microprocessor.

(8) *Lead angle change signal generator.* The position at which the lead angle is to be changed from one value to another is sent from the microprocessor to this block. When this position is reached, a signal is shot and sent to the microprocessor as an interrupt signal.

7.3.3 *Software*

There are lots of ideas for software. As a comparatively simple example, a program for driving a character wheel in a word processor is given here. The flow-chart is shown in Fig. 7.21, and the details are as follows.

(1) *Initial setting.* At first, the motor is operated in the ordinary open-loop mode to drive one revolution, or a little more, in the CW direction. The purpose of this is to reset the present position counter; when the reference signal, which is shot once a rotation, comes the counter is set to zero.

(2) *Pre-learning run.* In this process, every kind of motion which may be commanded in the job processing is executed several times to train the microprocessor system before it learns the timings to change lead angle to yield optimum speed patterns for a given load. Since the present application is to the drive of a 200-character wheel, motions of one to a hundred steps in both directions are instructed for learning the optimum timings to change lead angles. Two hundred is actually a large number of characters to be distributed on the periphery of a wheel with a low moment of inertia. In an actual machine with 128 or 132 characters, a double daisy wheel shown in Fig. 8.3 (p. 224) is used. For the sake of simple software, it is assumed here that the wheel is single row.

The learning method is explained in Figs 7.22 to 7.24, which all illustrate CW 13 step motions. The distance or number of steps to the new target is sent out from output port 01 to the error counter. Then (Accel, CW) mode of lead angle and direction is given to the logic sequencer from the output port 02 to start the motor. The motor is thus accelerated with a lead angle of 2.0 steps until it arrives at the switching

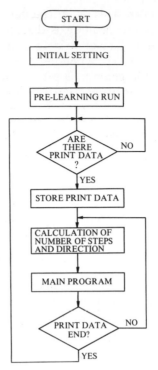

Fig. 7.21. Flow-chart of program.

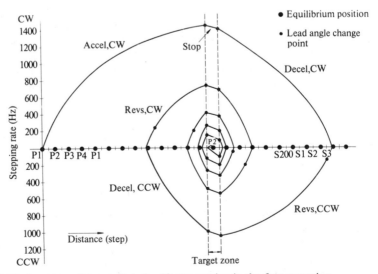

Fig. 7.22. Speed vs. distance plots for 13-step motion in the first execution.

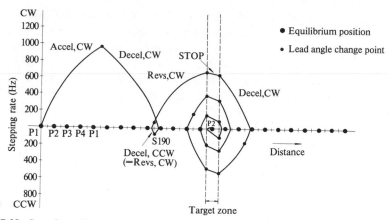

Fig. 7.23. Speed vs. distance plots for 13-step motion in the second execution.

point nearest to the target equilibrium point. When the rotor is in the target zone the lead angle is automatically set to the stop mode by the hardware. The motor begins to be decelerated. If the rotor exceeds the target zone and crosses the next switching point, an exceed interrupt signal is shot by the LACS (Lead Angle Change Signal) generator. Receiving this signal, the microprocessor commands a change in the lead angle to (Decel, CW) mode or −0.5-step mode to effectively decelerate the motion by the CCW torque (due to this mode of lead angle). The motor will eventually stop and start to move in the opposite direction. When the motor passes switching point S3 having reversed direction, a reverse interrupt signal is shot by the LACS generator. Receiving this signal, the microprocessor finds out the number of steps overshot, which is eight in Fig. 7.22. Note that the lead angle mode automatically becomes (Revs, CCW) or −1.5-step in the CCW direction after reversal of motional

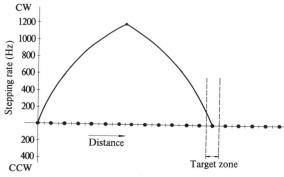

Fig. 7.24. Optimum speed vs. distance plots for 13-step motion.

direction. We can understand this using Table 7.1 as follows: suppose that the rotor position is, for example, in a P3 zone when the motor is driven in the (Decel, CW) mode. The phase which is now excited is found to be Phase 2 as underlined in the CW section of Table 7.1. On the other hand, in the CCW section of the same table, this number '2' appears, in the P3 column, in the (Revs, CCW) row for the lead angle of 1.5 steps. Hence, the motor will, after reversing the motion, be accelerated with a lead angle of 1.5 steps.

When the motor enters the target zone again, the lead angle is set to the stop mode to accommodate the rotor at the target equilibrium position. But a quick stop there may not be possible this time due to load inertia. If the target zone is again exceeded, another exceed interrupt signal is shot to change the lead angle to the (Decel, CCW) mode so as to decelerate and reverse the motion again. After recovery of the CW motion, the lead angle is automatically changed to the (Revs, CW) mode. The acceleration up to the target zone is not so good as before because of a smaller lead angle of 1.5 steps. The lead angle is set to zero again in the target zone by hardware. It is seen from the figure that another one or more cycles of oscillation will occur before the final accommodation at target.

When the same instruction, to move 13 steps in the CW direction, is given again, deceleration will be initiated at the switching point eight steps ahead of the target zone. In this case, the speed will become zero before arriving at the target shown in Fig. 7.23. When switching point S190 is crossed, a reverse interrupt is shot and the number of steps yet to be completed is counted by the error counter (four in the figure). This is regarded as a negative overshoot of four steps. The memory of overshoot is now corrected to $8 - 4 = 4$. Again in Fig. 7.23, the lead angle becomes that for the (Decel, CCW) mode, being the same as (Revs, CW) mode, at the interrupt signal to recover the CW movement. The execution after that is the same as before.

Repeating this process several times, the optimum timing for changing lead angle is found to attain the curve shown in Fig. 7.24.

(3) *Are there print data?* After 200 kinds of motions are trained, the system is ready to work for practical job processing. If the microprocessor receives any data to be printed, they are at once stored in RAMs.

(4) *Calculation of number of steps and direction.* The distance between the present position and the target is calculated, and the result is sent out to the error counter. The position at which lead angle is to be changed from accel to decel is also sent out to the LACS generator.

(5) *Main program.* Accel mode of lead angle and direction is set and put out from output port 2 to the gate to start the motor. It is expected that

every motion is performed in the shortest time. If, however, any variation of load conditions happens, overshoots positive or negative will occur, and the timings of changes in the lead angle are always corrected.

7.4   Other types of closed-loop control

7.4.1   *Closed-loop drive with current control*

Closed-loop control of a 132-step three-phase VR motor by means of current adjustment was developed by the Yokosuka Electrical Communication Laboratory of Nippon Telegraph and Telephone Public Corporation and Sanyo Denki Co., Ltd. to apply to a daisy wheel drive in a serial printer. Some details of this system are found in Ref. [5] in Japanese, but the outline of it is given here. The motor combined with an optical incremental encoder is shown in Fig. 7.25.

The block diagram of this system is shown in Fig. 7.26, and the features are as follows:

(i) The velocity profiles are fixed beforehand. The maximum distances are 66 steps in either direction. The speed profiles for them are stored in ROMs. The basic speed patterns are trapezoidal, but triangular patterns are used for short motions.

(ii) The position/speed is fed back, and the speed is compared with the command pattern. The error between them is amplified and used to control the current.

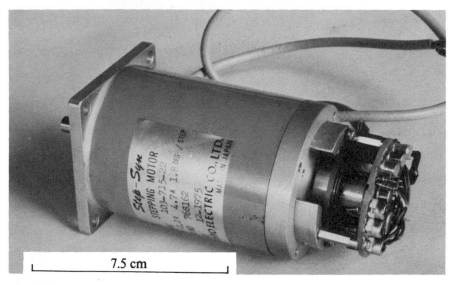

Fig. 7.25. A three-phase VR motor and an optical encoder.

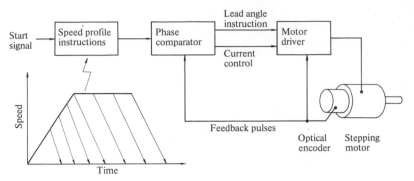

Fig. 7.26. Block diagram of closed-loop drive with current control.

(iii) The lead angles used are: 1.75 steps for acceleration, 1.25 steps for slewing and 0.5 step for deceleration.

(iv) The motor driver used is a chopper type.

(v) Step command signals during acceleration and slewing are taken from the encoder feedback signals, while the command signals are used during deceleration.

7.4.2 *Closed-loop control without position encoder*

The closed-loop control needs, basically, some type of encoder which translates mechanical positions into electrical signals. Encoders used for this purpose are, in many cases, of optical type. There are, however, some disadvantages and limitations of the optical encoder: firstly it is usually more expensive than the stepping motor itself; secondly the adjustment of the encoder position relative to the rotor's reference position is not an easy matter; and thirdly encoders are subject to environmental problems, such as heat or electrical noise.

Instead of using an optical encoder or another type of mechanical encoder to be coupled to the rotor, it is possible to detect the rotor position by current or voltage waveforms. Kuo[6] gave arguments on several ways of current sensing and a theory of closed-loop control of a VR stepping motor by current sensing.

In the 1979 international conference on stepping motors and systems, a novel scheme of closed-loop control of a hybrid linear motor, developed by Langley and Kidd,[7] was presented. The outline of this is given here. The cross-sectional view of the motor is shown in Fig. 2.49. The current waveforms in the winding just switched on are such as shown in Fig. 7.27(a), the upper trace represents drive without resultant motion and the lower trace represents the motion of the slider under no-load condition. If the motor is not allowed to reach full speed, for example, by having to work against the force of gravity or by being connected to an inertial

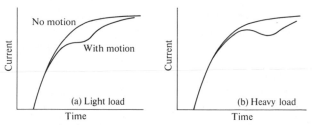

Fig. 7.27. Current waves in the winding just turned on: (a) no load; (b) loaded.

load, the lower waveform of Fig. 7.27(b) results. The feedback technique used with this linear motor is based on the difference between these waveforms.

Figure 7.28 shows a block diagram for this control scheme. The motor is excited in the bipolar two-phase-on mode, and each end of each winding is connected to a voltage comparator. The other comparator input is driven by a simple exponential-waveform generator, triggered by the microprocessor. After commanding a step, the microprocessor delays the exponential waveform so that it will occur entirely after the 'no motion' signal of Figs. 7.27(a) and (b). If there is no step motion there will be no crossover of the voltages, and ultimately no step confirmation is received by the microprocessor. However, if a 'motion waveform' such as those in Figs. 7.27(a) and (b) is received by the comparator, there will be two voltage crossings, switching the output of the comparator off and on again. It is reported that the timing of the first crossover is optimum for triggering the next step when the motor is to be accelerated or operating

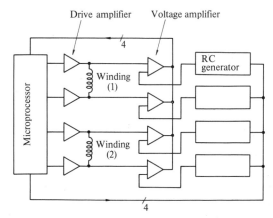

Fig. 7.28. Block diagram of closed-loop drive of a four-phase motor employing current-wave sensing.

in slewing. The second crossover can be used equally well for deceleration. The operation is done in the following manner:

(1) The first step in the desired direction is commanded by switching current in the appropriate winding.

(2) After a delay, the exponential-waveform generator is triggered by the microprocessor.

(3) When the first crossover is detected at the output of the comparator, the microprocessor determines how many more steps will be required to complete the motion. If the count remaining is more than two, the next drive state is selected and current immediately reversed in the proper windings for acceleration. If there are two or fewer steps remaining, the microprocessor sets a flag which delays this action until the second crossover. This delayed excitation of the winding makes use of the 'electromagnetic spring action' to effectively decelerate the motor.

References for Chapter 7

[1] Shimotani, K. and Kataoka, T. (1976). Improvement of a closed-loop stepping motor by excitation methods. *Trans. Institute of Electrical Engineers of Japan* **96B,** (6), 299–306.
[2] Kenjo, T. and Niimura, Y. (1979). *Fundamentals and applications of stepping motors.* (In Japanese.) p. 208. Sogo Electronics Publishing Co., Ltd., Tokyo.
[3] Fredriksen, T. R. (1968). Applications of the closed loop stepping motor. *IEEE Trans. Automatic Control* **AC13,** (5), 464–74.
[4] Kenjo, T., Takahashi, H., and Takahara, T. (1980). Microprocessor controlled self-optimization drive of a step motor. *Proc. Ninth annual symposium on Incremental motion control systems and devices.* Incremental Motion Control Systems Society, Champaign, Illinois, pp. 115–24.
[5] Kenjo, T. and Niimura, Y. (1979). *Fundamentals and applications of stepping motors.* (In Japanese.) p. 218. Sogo Electronics Publishing Co., Ltd., Tokyo.
[6] Kuo, B. C. (1979). Incremental motion control—Step motors and control systems. p. 256. SRL Publishing Company, Champaign, Illinois.
[7] Langley, L. W. and Kidd, H. K. (1979). Closed-loop operation of a linear stepping motor under microprocessor control. *Proc. International Conference on Stepping Motors and Systems.* University of Leeds, pp. 32–6.

8. Application of stepping motors

Some examples of stepping motor applications were cited in various parts of the preceding chapters. In this chapter we will have a look at applications of stepping motors from various viewpoints.

8.1 In computer peripherals

This field is considered to be the main area for stepping motor applications. There is a wide variety of computer peripherals. Let us consider a few major devices employing stepping motors.

8.1.1 Serial printers

A serial printer is an automatic typewriter used in a word processor system, and is a device that prints out data one character at a time. An external view and the internal mechanism of the device are shown in Fig. 8.1. The construction of the serial printer differs slightly from manufacturer to manufacturer. The basic structure of the character impact type is, however, such as shown in Fig. 8.2. A character wheel is coupled directly to the motor mounted in the carriage. Several varieties of character wheels are in use: a daisy wheel is shown in Fig. 2.7 on p. 21; a badminton ball type and a double daisy wheel are shown in Fig. 8.3. All of them are made of light materials like plastic. The number of characters is, for example, 64, 96, 128, or 132. The last two numbers are employed in the character wheels which have, for example, Roman and Japanese alphabets. In general, a stepping motor is used for driving the character wheel in the open-loop control mode for the applications with printing speeds as slow as 30 characters per second or so. For high speed applications a DC servo-motor is used in the feedback control scheme. It is also possible to drive a stepping motor in the closed-loop mode for a high-speed application. A stepping motor or a DC servo-motor is used for carriage transport. PM stepping motors of 45° or 90° angle step are employed for feeding the ribbon. A stepping motor is also used to drive the platen which feeds the paper.

8.1.2 Application of linear motors to printers

The principle of the application of a VR linear stepping motor[1] to the carriage transport in a serial printer was illustrated in Fig. 1.13. An actual machine based on this idea is shown in Fig. 8.4. The carriage provided

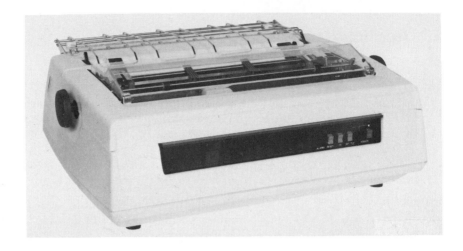

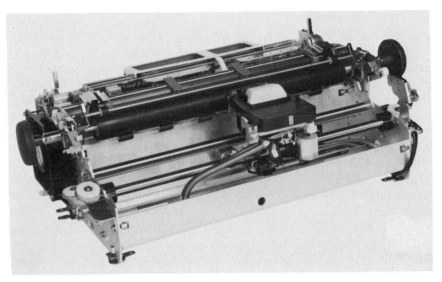

Fig. 8.1. External view and internal mechanism of a serial printer. (By courtesy of Nippon Electric Co., Ltd.)

with a three-phase winding acts as a motor or slider. The rail or stationary base is made of laminated silicon steel and teeth are provided at the top and bottom at the same pitch.

Figure 8.5[2] shows another VR linear motor applied to the carriage transport. In this machine, however, the slider has no windings, while many coils are provided on the stator.

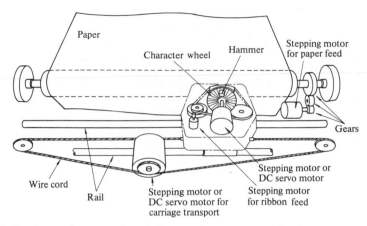

Fig. 8.2. Fundamental construction of character-impact type serial printer.

8.1.3 XY-plotters

As explained in Chapter 1, stepping motors are used for driving the pen in an XY-plotter or graph plotter, which is a device to produce a permanent graphical representation of computational results. Figure 8.6 shows a recent XY-plotter and Fig. 8.7 illustrates the pen drive system[3] employed in this plotter for one axis. In the plotter shown in Fig. 1.12 on p. 10, gears are used to transmit the motion of the motor shaft to the pen. In the system of Fig. 8.7, however, nylon-coated stainless steel fibre cable is used, which is advantageous in the fact that there is no mechanical play and audible noise is low. To have good line quality and high resolution, the ministep drive explained in Section 2.3.6[4] is employed in the plotter of Fig. 8.6.

8.1.4 Floppy disc drives

The floppy disc drive is used as an auxiliary memory for every class of computer, from micro to large. It is also used as a memory device in an intelligent terminal. The floppy disc, known also as the flexible disc or disket, is a mylar film disc, one (or both) side(s) of which is coated with magnetic material in which data are stored. As shown in Fig. 8.8 the disc is enclosed in a plastic jacket, and it is mounted on a driving unit with a speed of 300 or 360 r.p.m. When mounted on the driving unit, only the disc is coupled to the driving mechanism, while the jacket remains stationary. Various types of synchronous motors and brushless DC motors are in use as the driving motors. As will be shown soon a stepping motor is used to drive the magnetic head. It is known that a stepping motor can also be used as the disc drive motor.

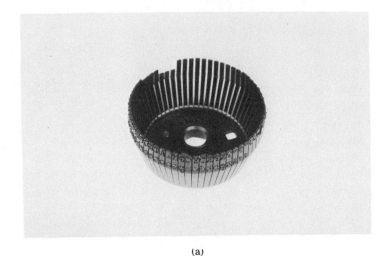

(a)

(b)

Fig. 8.3. Character wheels: (a) badminton ball type; and (b) double daisy wheel.

A disc can store 250 K to 1 M bytes of data. Reading and writing of data are carried out by one or two magnetic heads. The track pitch is 0.5 mm and the number of tracks is 77 or more in the full-size disc, and 35 or more in the mini-size one. The diameters of the full and mini-size discs are 20.3 cm (=8 inches) and 13.3 cm (=$5\frac{1}{4}$ inches), respectively. Figure 8.9 illustrates the drive mechanism of a one-sided floppy disc. The head assembly is carried between tracks by a stepping motor and lead

Fig. 8.4. A VR linear motor designed for carriage transport in a serial printer. (By courtesy of IBM Corporation.)

screw. The most commonly used motor in the full-size system is a three- or four-phase VR motor of 15° step (see Fig. 8.10). The head carriage moves one track pitch when the motor advances by 15°. In recent years, 1.8° hybrid motors have been used to drive two heads in the drives of double-sided floppy discs, in which heads are driven by a stainless steel belt. Claw-pole or can-stack types of stepping motors are extensively employed in the mini-size floppy drives.

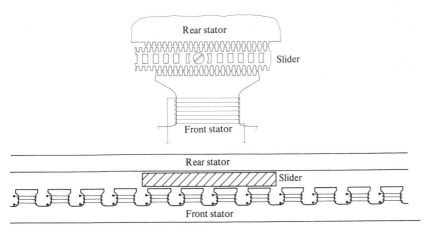

Fig. 8.5. A VR linear motor employed in a printer. (After Ref. [2], reproduced by permission of Professor P. J. Lawrenson and Dr G. Singh)

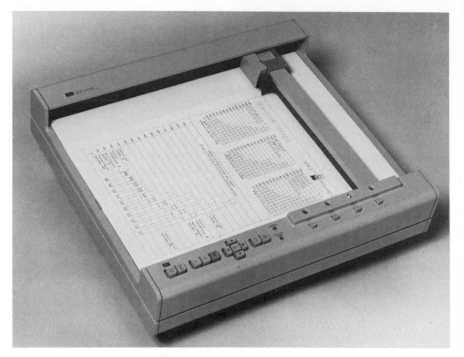

Fig. 8.6. Hewlett Packard HP9872B graph plotter. (By courtesy of Hewlett Packard.)

8.2 Applications in numerical control

Another big application field for stepping motors is found in the numerical controls of machine tools and workpieces, etc.

8.2.1 XY-tables and index tables

A device for controlling position on a single plane by using two motors to control the X and Y movement is called an XY table. For this, stepping

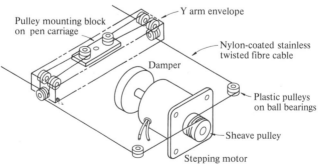

Fig. 8.7. Driving mechanism of XY-plotter using two stepping motors. (After Ref. [3], reproduced by permission of Mr M. L. Patterson.)

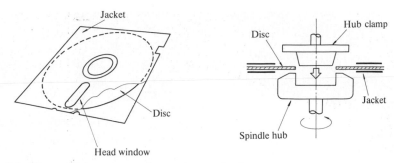

Fig. 8.8. Floppy disc structure and mounting mechanism.

motors are used as shown in Fig. 8.11(a). A numerical index table can be controlled by a stepping motor too (see Fig. 8.11(b)). In these devices, machining is done after positioning or indexing is completed.

8.2.2 Milling machines

The three-axis movement of the workpiece in a numerical milling machine can be controlled by three stepping motors. In Fig. 8.12 the two motors to govern the X- and Y-axes are illustrated, but the third motor which is installed below the table to govern the Z-axis is not drawn. In the numerical milling machine, the machining is done by the cutting tool while the motors are operated. Since, as discussed in Chapter 4, stepping

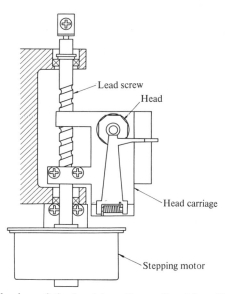

Fig. 8.9. Driving mechanism of the head in a floppy disc drive. (By courtesy of TEAC Corporation.)

Fig. 8.10. A VR motor for driving the head in a floppy disc drive. (By courtesy of Moore Reed and Co. Ltd.)

motors produce an oscillatory torque, the surface of the workpiece finished by the continuous path control using stepping motors is in general not so good as the surface finished by a DC servomotor operated NC machine.

8.2.3 *Automated drafting machine driven by a linear motor*

Stepping motors are utilized for the drive of the draft head in automated drafting systems. The principle is very similar to that of an XY-plotter, but the size of plane is larger than those in XY-plotters. Drafting machines are usually controlled by a dedicated mini-computer. Here we

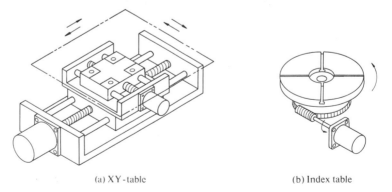

(a) XY-table   (b) Index table

Fig. 8.11. Numerically controlled XY-table and index table.

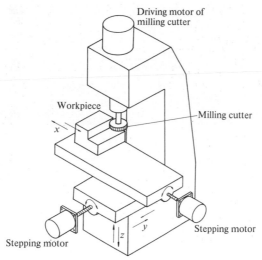

Fig. 8.12. Principle of numerically controlled milling machine using three stepping motors.

will look at a drafting machine in which the pen is driven by a planar motor which is a combination of two linear stepping motors.

The basic construction and principle of the Sawyer stepping motor were discussed in Section 2.2.7. By combining two linear motors in the way shown in Fig. 8.13,[5] one can realize a motor which can travel in any direction on a stationary base; one motor is used to produce force in the

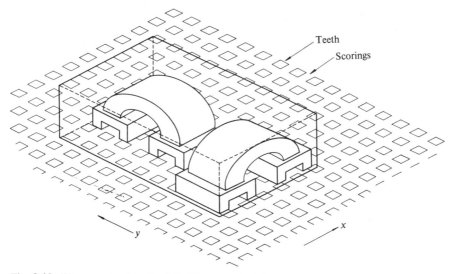

Fig. 8.13. Planar motor based on the Sawyer principle.

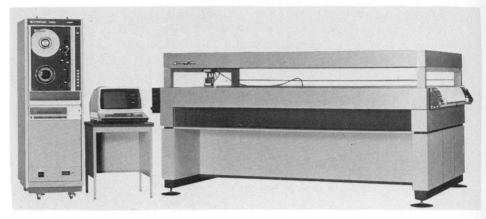

Fig. 8.14. Automated drafting system. (By courtesy of Daini Seikosha Co., Ltd.)

Fig. 8.15. Xynetics drafting head. (By courtesy of Daini Seikosha Co., Ltd.)

X-direction, and other in the Y-direction. The pictures of a drafting system and the drafting head are shown in Figs. 8.14 and 8.15, respectively. One linear motor assembled in the head has the construction of Fig. 2.50 (p. 48), with a tooth pitch of 0.96 mm. The stationary base is cross-scored to form equidistant square islands in a waffle pattern. The scorings are filled with non-magnetic material to provide a smooth surface so that the motor can float above or below the stationary base on a stable airbearing. As shown in Fig. 8.14 the drafting head, being about 1.5 kg, is suspended from the stationary base by the magnetic force, but free to move due to a thin air film of about 10 μm. Air is supplied from four holes to the surface. The motor is driven in the mini-step mode; the exciting current is a 96-stepped sinusoidal wave. Consequently the resolution is 960 μm/96 = 10 μm. In this drafting system the motor is driven in the open-loop mode, and the maximum speed of 1 m s^{-1} and an acceleration of 9.8 m s^{-2} are attained.

To be free from the characteristic oscillatory phenomena in the open-loop mode, three accelerometers are mounted in the head to detect oscillations and feedback to the driver amplifier to suppress undesirable motions. Two accelerometers are used for detecting acceleration in the X- and Y-directions, and the third one is necessary for detecting rotational oscillation.

8.2.4 Tape punch and tape reader

The use of stepping motors for the drive of the sprocket in a tape punch and a tape reader (Fig. 8.16) was suggested at the beginning of

Fig. 8.16. A paper tape reader. (By courtesy of Japan Macnics Co., Ltd.)

Chapter 2. Tape punches with speeds as high as 100 to 200 lines per second use stepping motors for the drive of their sprockets. For low-speed devices a ratchet mechanism is used. The read speed in the tape reader ranges from 200 to 500 lines per second.

8.2.5   *Tape trace checker*

The tape trace checker is used to check the data on the tape programmed for NC lathes. A stepping motor is used to drive the tape sprocket. An external view of the device is given in Fig. 8.17. The machining process is drawn on the recording paper by the pen attached to the device (see Fig. 8.18).

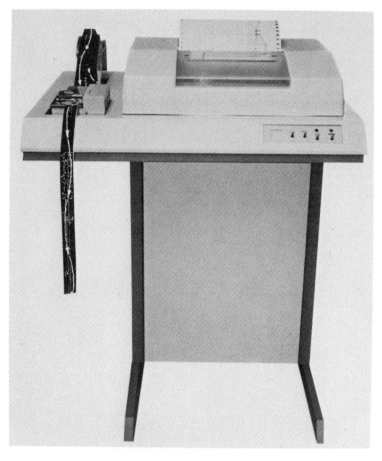

Fig. 8.17. Tape trace checker. (By courtesy of Ricoh Co., Ltd.)

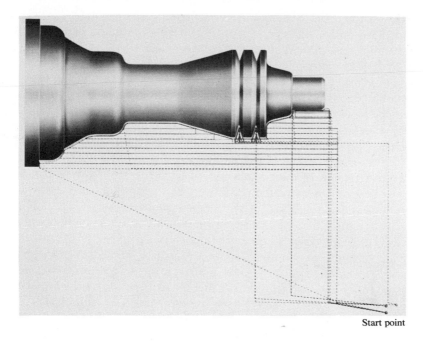

Start point

Fig. 8.18. Relationship between NC lathe finished product and the blade trace. (By courtesy of Ricoh Co., Ltd.)

8.3 Other applications

(1) *Facsimiles.* The device designed to transmit documents, drawings, etc. to distant locations via telephone lines is called a facsimile machine. Facsimile machines are becoming popular for sending copies of documents to overseas countries, as well as for local transmissions. The basic principle of a facsimile machine is shown in Fig. 8.19. The document wrapped on the drum is scanned in horizontal (main) and rotating (sub) directions. The document is divided into graphic elements which are converted into electrical signals by a photoelectric reading head. The

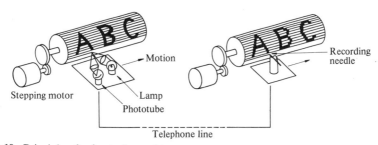

Fig. 8.19. Principle of a facsimile machine.

signals are then sent over public lines to the receiving unit. The signals received are demodulated and reproduced by a recording pen. Main and sub-scannings are repeated on the receiving side.

Synchronous motors (hysteresis motors or reluctance motors) or stepping motors are used to drive the drum for the sub-scanning in both the transmitting and receiving units. Synchronous motors are used in facsimile machines in which the drums in both units must rotate at the same speed, synchronizing with each other. In more advanced facsimile machines, the documents are first scanned, the data thus obtained are stored temporarily in the buffer memory, the data is then compressed to reduce the amount of data to the necessary level, and lastly it is sent to the temporary memory of the receiving side. The motor used in this class of machine is a stepping motor. If sufficient memory capacity is available, a large amount of the document can be transmitted in a single operation. However this increase in memory increases the cost of the machine. Instead, storing and transmitting are repeated using a small capacity memory. Consequently the motor is driven in the incremental fashion. The amount of data transmitted per increment depends on the type of the document, memory capacity, compressor, and line conditions. The number of scannings per mm is, for example, four or eight. The greater this number, the better the quality of reproduced documents or drawings.

(2) *Semi-automatic wiring unit.* Figure 8.20 illustrates a semi-automatic wiring machine for printed circuit boards and its control panel. The box on the left side stores lead wires of different lengths and colours. The

Fig. 8.20. Semi-automatic wiring machine and control panel. (By courtesy of Sanyo Denki Co., Ltd.)

control panel on the right side incorporates a tape reader. All the wiring procedures are programmed and stored on the tape. As soon as an instruction is issued to start wiring by pressing a button the wire required first is indicated by a lamp. Simultaneously the two stepping motors for driving the vertical and horizontal axes are activated and the hole for inserting the wire wrapper is specified. The wrapping must be done manually in this semi-automatic system. Both motors are activated as soon as the device is notified that wrapping has been completed, which is done by pressing a foot pedal. They will specify the point to which the other end of the wire is to be connected. As soon as a completion signal is issued, the second wire is specified by a lamp, at the same time the motors are driven to the point of wrapping.

(3) *Applications in space science.* Stepping motors are used in spacecraft launched for scientific exploration of planets. The case of Mariner Jupiter Saturn 77 (or Voyager 1) was reported in References [6] and [7]. The outlines of the stepping motors used in it are reproduced here. Various scientific instruments such as TV cameras and ultraviolet spectrometers are mounted on the scientific scan platform. Stepping motors are used to point the instruments at desired targets. The motor used in this spacecraft is an 11 size (about 27 mm in diameter) 90-degree, four-phase motor of permanent magnet type. The motor gear arrangement of 9081:1 reduction ratio illustrated in Fig. 8.21, which is actually enclosed in a can-type container, provides 0.1729 mrad (about 0.01 degree) of shaft rotation for each motor pulse. The feedback and telemetry potentiometers, which are connected to the actuator output shaft by means of high-precision anti-backlash gearing, give an accurate measure of the scan actuator shaft position.

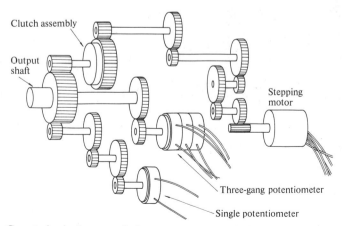

Fig. 8.21. Gear trains in the scan platform.

Fig. 8.22. Rubber stamp printer used to mark steel plates.

(4) *Character printing board.* This is an application in heavy industry. The device shown in the photograph of Fig. 8.22 is used to print the production number, etc. on the steel plates manufactured in a steel mill. A stepping motor is used to drive a rubber stamp, and a number of the devices are simultaneously operated by a controller.

Fig. 8.23. Digital-to-analogue converter employing a stepping motor. (By courtesy of Sanyo Denki Co., Ltd.)

(5) *Mechanical digital-to-analogue converter.* Stepping motors can be used in electromechanical D–A converters. The device shown in Fig. 8.23 employs potentiometers for the analogue section. A potentiometer is coupled to the shaft of a stepping motor via gears. The digital signal applied to the motor drive is converted into an analogue voltage at the output terminals of the potentiometer. Synchros can also be used in the analogue part.

(6) *Electron-beam microfabricator.* In the process of manufacturing a large-scale integrated circuit, two stepping motors are used to drive a precision XY-table carrying a wafer in a vacuum chamber while the wafer is exposed to an electron beam. Figure 8.24 shows the electron optics and workstage area. Two stepping motors are seen.

Fig. 8.24. Electron-beam microfabricator. (By courtesy of Cambridge Instruments.)

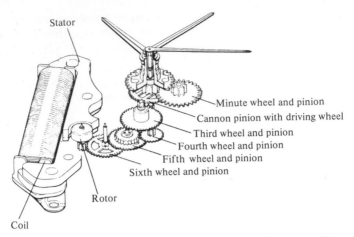

Fig. 8.25. Principle of the watch employing a single-phase stepping motor. (By courtesy of Citizen Watch Co., Ltd.)

(7) *Application in watches*. A very tiny single-phase stepping motor is used in a watch as shown in Fig. 8.25. The principle of this motor is discussed in Section 2.4.1. In typical watches, highly stable oscillation of 32 768 Hz or 2^{15} Hz, produced by a quartz crystal oscillator, is divided down to 1 Hz through a divider circuit. The signal thus divided to 1 Hz is amplified enough to energize the stator coil to drive the rotor at the rate of 1/2 revolution per second. The rotor is a samarium cobalt magnet of about 1.5 mm diameter. The crystal oscillator and electronic circuit are packaged in a very small space in a watch.

References for Chapter 8

[1] Chai, H. D. and Pawletko, J. P. (1977). Serial printer with linear motor drive. U.S. Patent 4,044,881.
[2] Singh, G., Gerner, M., and Itzkwitz, H. (1979). Motion control aspects in the Qyx intelligent typewriter. *Proc. International Conference on Stepping Motors and Systems*. University of Leeds. pp. 6–12.
[3] Patterson, M. L., Haselby, R. D., and Kemplin, R. M. (1977). Speed, precision, and smoothness characterize four-color plotter pen drive system. *Hewlett Packard Journal* **29**, (1), 13–19.
[4] Patterson, M. L. and Haselby, R. D. (1977). A micro-stepped XY controller with adjustable phase current waveforms. *Proc. Sixth Annual Symposium on Incremental motion control systems and devices*. Department of Electrical Engineering, University of Illinois. pp. 163–8.
[5] Hinds, W. E. and Nocito, B. (1973). The Sawyer linear motor. *Proc. Second Annual Symposium on Incremental motion control systems and devices*. Department of Electrical Engineering, University of Illinois. pp. W1-10.

[6] Hughes, R. O. (1975). Dynamics of incremental motion devices associated with planetary exploration spacecraft. *Proc. Fourth Annual Symposium on Incremental motion control systems and devices.* Department of Electrical Engineering, University of Illinois. pp. BB 1–8.

[7] Tolivar, A. F. and Hughes, R. O. (1976). Science platform pointing control law for a planetary exploration spacecraft. *Proc. Fifth Annual Symposium on Incremental motion control systems and devices.* Department of Electrical Engineering, University of Illinois. pp. AA1–2.

Index